THE WITNESS OF THE STARS

THE
WITNESS OF THE STARS

BY THE

REV. ETHELBERT W. BULLINGER, D.D.

"HE telleth the number of the stars;
He giveth them all their names."

(Ps. Cxlvii. 4. R,V.)

GYAN

ISBN : 978-81-212-9012-8 (HB)

First Published, 1893

Indian Reprintg in 2017

Published by

Gyan Publishing House
5, Ansari Road
Daryaganj, New Delhi-110002
Phone: 9811692060
E-mail: books@gyanbooks.com

Printed at: G. Print Process, Delhi.

PREFACE.

Some years ago it was my privilege to enjoy the acquaintance of Miss Frances Rolleston, of Keswick, and to carry on a correspondence with her with respect to her work, *Mazzaroth: or, the Constellations.* She was the first to create an interest in this important subject. Since then Dr. Seiss, of Philadelphia, has endeavoured to popularize her work on the other side of the Atlantic; and brief references have been made to the subject in such books as *Moses and Geology*, by Dr. Kinns, and in *Primeval Man;* but it was felt, for many reasons, that it was desirable to make another effort to set forth, in a more complete form, the *witness of the stars to prophetic truth*, so necessary in these last days.

To the late Miss Rolleston, however, belongs the honour of collecting a mass of information bearing on this subject; but, published as it was, chiefly in the form of *notes*, unarranged and unindexed, it was suited only for, but was most valuable to, the student. She it was who performed the drudgery of collecting the facts presented by Albumazer, the Arab astronomer to the Caliphs of Grenada, 850 A.D.; and the Tables drawn up by Ulugh Beigh, the Tartar prince and

a 2

astronomer, about 1450 A.D., who gives the Arabian Astronomy as it had come down from the earliest times.

Modern astronomers have preserved, and still have in common use, the ancient names of over a hundred of the principal stars which have been handed down; but now these names are used merely as a convenience, and without any reference to their significance.

This work is an attempt to popularize this ancient information, and to use it in the interests of truth.

For the ancient astronomical facts and the names, with their signification, I am, from the very nature of the case, indebted, of course, to all who have preserved, collected, and handed them down; but for their interpretation I am alone responsible.

It is for the readers to judge how far my conclusions are borne out by the evidence; and how far the foundation of our hopes of coming glory are strengthened by the prophecies which have been written in the stars of heaven, as well as in the Scriptures of truth.

For the illustrations I am greatly indebted to Jamieson's *Celestial Atlas*, 1820; Flammarion's *L'Etoiles;* Sir John W. Lubbock's *Stars in Six Maps*, 1883; and to the late Mr. Edward J. Cooper's *Egyptian Scenery*, 1820. For the general presentation and arrangement of the Constellations I am responsible, while for the drawings my thanks are due to my friend Miss Amy Manson.

It is the possession of "that blessed hope" of Christ's speedy return from Heaven which will give true interest in the great subject of this book.

No one can dispute the antiquity of the Signs of the Zodiac, or of the Constellations. No one can question the accuracy of the ancient star-names which have come down to us, for they are still preserved in every good celestial atlas. And we hope that no one will be able to resist the cumulative evidence that, apart from God's grace in Christ there is no hope for sinners now: and apart from God's glory, as it will be manifested in the return of Christ from Heaven, there is no hope for the Church, no hope for Israel, no hope for the world, no hope for a groaning creation. In spite of all the vaunted promises of a religious World, and of a worldly Church, to remove the effects of the curse by a Social Gospel of Sanitation, we are more and more shut up to the prophecy of Gen. iii. 15, which we wait and long to see fulfilled in Christ as our only hope. This is beautifully expressed by the late Dr. William Leask:—

And is there none before? No perfect peace
Unbroken by the storms and cares of life,
Until the time of waiting for Him cease,
By His appearing to destroy the strife?
No, none before.

Do we not hear that through the flag of grace
  By faithful messengers of God unfurled,
All men will be converted, and the place
  Of man's rebellion be a holy world?
                    Yes, so we hear.

Is it not true that to the Church is given
  The holy honour of dispelling night,
And bringing back the human race to heaven,
  By kindling everywhere the Gospel light?
                    It is not true.

Is this the hope—that Christ the Lord will come,
  In all the glory of His royal right,
Redeemer and Avenger, taking home
  His saints, and crushing the usurper's might?
                    This is the hope.

May the God of all grace accept and bless this effort to show forth His glory, and use it to strengthen His people in waiting for His Son from Heaven, even Jesus which delivered us from the wrath to come.

ETHELBERT W. BULLINGER.

August 31st, 1893.

LIST OF ILLUSTRATIONS.

*** All are drawn to the same *relative* scale, except Plates VI., VIII., XIV., and XXV., which are enlarged; while Plates III., XI., XVIII., XIX., XX., XXI., XXVI., XXX., and XL., are slightly reduced.

ENGRAVINGS.

THE

WITNESS OF THE STARS.

FOR more than two thousand five hundred years the world was without a written revelation from God. The question is, Did God leave Himself without a witness? The question is answered very positively by the written Word that He did not. In Rom. i. 19 it is declared that, "that which may be known of God is manifest in them; for God hath showed it unto them. For the invisible things of Him from the creation of the world are clearly seen, being understood by the things that are made, even His eternal power and Godhead; so that they are without excuse." But how was God known? How were His "invisible things," *i.e.*, His plans, His purposes, and His counsels, known since the creation of the world? We are told by the Holy Spirit in Rom. x. 18. Having stated in *v.* 17 that "Faith cometh by hearing and hearing by the Word (ῥῆμα, *the thing spoken, sayings*) of God," He asks, "But I say, Have they not heard? Yes, verily." And we may ask, How have they heard? The answer follows—"Their sound went into all the earth (γῆ) and their words (ῥήματα, *their teaching, message, instruction*) unto the ends of the world (οἰκουμένη)." What words? What instruction? Whose message? Whose teaching? There is only one answer, and that is, THE

S. 1005. A

HEAVENS! This is settled by the fact that the passage is quoted from Ps. xix., the first part of which is occupied with the Revelation of God written in *the Heavens*, and the latter part with the Revelation of God written in the *Word:*

This is the simple explanation of this beautiful Psalm. This is why its two subjects are brought together. It has often perplexed many why there should be that abrupt departure in verse 7—"The law of the LORD is perfect, converting the soul." The fact is, there is nothing abrupt in it, and it is no departure. It is simply the transition to the second of the two great Revelations which are thus placed in juxtaposition. The first is the Revelation of the Creator, *El*, אֵל, in His *works*, while the second is the Revelation of the Covenant Jehovah, יְהוָה, in His *Word*. And it is noteworthy that while in the first half of the Psalm, *El* is named only once, in the latter half *Jehovah* is named *seven* times, the last being threefold (Jehovah, Rock, and Redeemer), concluding the Psalm.

Let us then turn to Ps. xix., and note first—

The Structure of the Psalm as a whole.*

A | 1–4–. The Heavens.
  B | –4–6. "In them" (בָּהֶם) the Sun.
A | 7–10. The Scriptures.
  B | 11–14. "In them" (בָּהֶם) Thy Servant.

---

* For what is meant by "Structure," see *A Key to the Psalms*, by the late Rev. Thos. Boys, edited by the present author, 7, St. Paul's Churchyard. Price Five shillings.

In the *Key to the Psalms*, p. 17, it is pointed out that the terms employed in *A* and *B* are *astronomical*,* while in A and B they are *literary*. Thus the two parts are significantly connected and united.

Ewald and others imagine that this Psalm is made up of two fragments of separate Psalms composed at different periods and brought together by a later editor!

But this is disproved not only by what has been said concerning the structure of the Psalm as a whole, and the interlacing of the astronomical and the literary terms in the two parts, but it is also shown by more minute details.

Each half consists of two portions which correspond the one to the other, A answering to *A*, and B to *B*. Moreover, each half, as well as each corresponding member, consists of the same number of lines ; those in

---

* *Viz.*, in *A* (verses 7, 8),—

"Converting," from שׁוּב, *to return*, as the sun in the heavens.

"Testimony," from עוּד, *to repeat*, hence, *a witness*, spoken of the sun in Ps. lxxxix. 37.

"Sure," אָמֵן, *faithful*, as the sun. (Ps. lxxxix. 37.)

"Enlightening," from אוֹר, *to give light*, as the sun. (Gen. i. 15, 17, 18 ; Isa. lx. 19 ; Ezek. xxxii. 7.)

In *B* (verses 11, 12, 13),—

"Warned," from זָהַר, *to make light*, hence, *to teach*, *admonish*.

"Keeping," from שָׁמַר, *to keep*, *observe*, as the heavens. (Ps. cxxx. 6 ; Isa. xxi. 11.) Or as the heavenly bodies *observe* God's ordinances.

"Errors," from שָׁגָה, *to wander*, as the planets.

"Keep back," חָשַׂךְ, *to hold back*, *restrain*.

"Have dominion over," from מָשַׁל, *to rule*. Spoken of the sun and moon in Gen. i. 18. "The sun to rule the day," &c. (Ps. cxxxvi. 8, 9.)

A 2

the first half being, by the *cæsura*, short, while those in the last half are long (or double).

A | 1–4–. Eight lines } 14 lines.
B | –4–6. Six lines }
A | 7–10. Eight lines } 14 lines.
B | 11–14. Six lines }

If we confine ourselves to the first half of the Psalm* (A and B, verses 1–6), with which we are now alone concerned, we see a still more minute proof of Divine order and perfection.

The Structure of A *and* B.

A & B | C | 1. The heavens.
D | 2. Their testimony: incessant. (Pos.)
E | 3. Their words inaudible. (Neg.)
D | 4–. Their testimony: universal. (Pos.)
C | –4–6. The heavens.

Here we have an *introversion*, in which the extremes (C and *C*) are occupied with the *heavens;* while the means (D, E and *D*) are occupied with their testimony.

The following is the full expansion of the above, with original emendations which preserve the *order* of the Hebrew words and thus indicate the nature of the structure:—

---

* The other half of the Psalm is just as perfectly arranged. For example, there are six words used (verses 7–9) to describe the fulness of the Word of God, and they are thus placed, alternately :—

F | *Two* feminine singulars. (Law and Testimony.)
G | *One* masculine plural. (Statutes.)
F | *Two* feminine singulars. (Commandment and Fear.)
G | One masculine plural. (Judgments.)

C | a | The heavens
b | are telling *
c | the glory † of God:
c | and the work of his hands
b | is setting forth ‡
a | the firmament.

D | d | Day after day §
e | uttereth || speech,
d | And night after night
e | sheweth knowledge.

E | f | There is no speech (what is articulate)
g | and there are no words; } what is audible
g | and without being audible, }
f | is their voice (what is articulate).

D | h | Into all the earth (as created)
i | is their line ¶ gone forth;
h | And into the ends of the world (as inhabited)
i | Their sayings.

C | j | For the sun He hath set a tent (an abode) in them;
k | l | and he as a bridegroom (comparison)
m | is going forth from his canopy, (motion: its rising)
l | he rejoiceth as a mighty one (comparison)
m | to run his course. (Motion: its rapid course.)
k | n | From the end of the heavens } egress
o | is his going forth, }
o | and his revolution } regress
n | unto their ends: }
j | and there is nothing hid from his heat (*i.e.*, from him **).

---

* From סָר, *to cut into*, or *grave*, hence, *to write*. It has the two senses of our English verb *tell*, which means *to count*, and also *to narrate*. The first occurrence is Gen. xv. 5, "*Tell* (סְפֹר) the stars, if thou be able to *number* (לִסְפֹּר) them." Gen. xxiv. 66, "The servant *told* Isaac all things that he had done." Ps. lxxi. 15, "My mouth shall *show forth* (יְסַפֵּר, *tell of*, R.V.) thy righteousness and thy salvation all the day; for I know not the *numbers* (סְפֹרוֹת, *i.e.*, *the accounts*) of them," *i.e.*, all the particulars.

Surely there is something more referred to here than a mere wonder excited by the works of the Creator! When we read the whole passage and mark its structure, and note the words employed, we are emphatically told that the heavens contain a revelation from God; they prophesy, they show knowledge, they tell of God's glory, and set forth His purposes and counsels.

It is a remarkable fact that it is in the Book of Job, which is generally allowed to be the oldest book in the

---

† From כָּבֵד, *to be heavy, weight,* the context determining whether the weight spoken of is advantageous or not. The first occurrence is Gen. xii. 10, "The famine was *grievous* (כָּבֵד) in the land." The next, xiii. 2, "Abram was very *rich* (כָּבֵד)." It is often applied to persons who are *of weight* and *importance*, hence, glorious and honourable. It is used of the *glory* of the Lord, and of God Himself, as we use Majesty of a person. See Isa. iii. 8; iv. 2; xi. 10; xliii. 20; Hag. ii. 8; Ex. xvi. 7; xxiv. 17; 1 Sam. iv. 21; Pss. xxvi. 8 (*honour*); lxiii. 3.

‡ From נָגַד, *to set before, to set forth, to shew.* First occurrence, Gen. iii. 11, "Who *told* thee that thou wast naked." Ps. xcvii. 6, "The heavens *declare* His righteousness"; cxi. 6, "*He hath shewed* his people the power of his works."

§ This is the English idiom for the Hebrew "Day to day." The ל is used in its sense of *adding* or superadding to, as in Isa. xxviii. 10, צַו לָצָו, "precept to precept;" *i.e.*, precept after precept, line after line. Gen. xlvi. 26, "All the souls that came with Jacob" (לְיַעֲקֹב, to Jacob; *i.e.*, in addition to Jacob. So here, "Day to day;" *i.e.*, Day in addition to day, or, as we say, Day after day).

‖ From נָבַע, *to tell forth*, akin to נָבָא, *to prophesy*, from root *to pour forth.* Lit., here, poureth forth discourse. Ps. cxlv. 9, "abundantly utter."

¶ Their line, קָו, *i.e.*, their measuring line. By the figure of metonymy the *line* which measures is put for the portion or heritage which is measured, as in many other places. See Ps. xvi. 6, "The lines are fallen unto me in pleasant places; yea, I have a goodly heritage." (See also Ps. lxxviii. 55, &c.) Here, it means that "Their measuring line has gone forth unto all the earth (אֶרֶץ)"; *i.e.*, All the earth inherits this their testimony (*i.e.*, has this testimony for its heritage), and to the ends of the world (תֵּבֵל, *the inhabited world*) their instruction has gone forth. With this agrees, in sense, the LXX. here, and Rom. x. 16, which each has φθόγγος, *a sound*, or *voice*; *i.e.*, a sound in relation to the hearer, rather than to that which causes it. The meaning of the passage is, "All the earth has their *sound* or testimony as its heritage, and the ends of the world hear their words." Symmachus has ἦχος, *a sound*, or *report.*

** חַמָּה means *that which is hot*, and is a poetical name of the sun itself.

Bible,* if not in the world, that we have references to this Stellar Revelation. This would be at least 2,000 years before Christ. In that book the signs of the Zodiac and the names of several stars and constellations are mentioned, as being ancient and well-known.

In Isa. xl. 26 (R.V.) we read :—

"Lift up your eyes on high,
And see who hath created these,
That bringeth out their host by number:
He calleth them all by name;
By the greatness of His might,
And for that He is strong in power,
Not one is lacking."

We have the same evidence in Psalm cxlvii. 4. (R.V.)

"He telleth the number of the stars;
He giveth them all their names."

Here is a distinct and Divine declaration that the great Creator both *numbered* as well as *named* the stars of Heaven.

The question is, Has he revealed any of these names? Have any of them been handed down to us?

The answer is Yes; and that in the Bible itself we have the names (so ancient that their meaning is a little obscure) of *Ash* (עָשׁ, a name still connected with the Great Bear), *Cesil* (כְּסִיל), and *Cimah* (כִּימָה).

They occur in Job ix. 9: "Which maketh Arcturus (R.V. *the Bear*), Orion, and Pleiades, and the chambers of the south." (Marg., Heb., *Ash*, *Cesil*, *and Cimah.*)

Job xxxviii. 31, 32: "Canst thou bind the sweet influences (R.V. cluster) of the Pleiades (marg., *the*

* Job is thought by some to be the Jobab mentioned in Gen. x. 29, the third in descent from Eber.

seven stars, Heb. *Cimah*), or loose the bands of Orion (marg. Heb. *Cesil*)? Canst thou bring forth Mazzaroth (marg., *the twelve signs*. R.V., "the twelve signs": and marg., *the signs of the Zodiac*) in his season? or canst thou guide Arcturus with his sons (R.V., the Bear with her train; and marg., Heb., *sons*)." *

Isa. xiii. 10: . . . "The stars of heaven and the constellations thereof." . . .

Amos v. 8: "Seek him that maketh the seven stars (R.V., the Pleiades) and Orion."

Then we have the term "Mazzaroth," Job xxxviii. 32, and "Mazzaloth," 2 Kings xxiii. 5. The former in both versions is referred to the Twelve Signs of the Zodiac, while the latter is rendered "planets," and in margin, *the twelve signs or constellations*.

Others are referred to by name. The sign of "Gemini," or the Twins, is given as the name of a ship: Acts xxviii. 11, Διόσκουροι (*i.e.* Castor & Pollux).

Most commentators agree that the constellation of "Draco," or the Dragon (between the Great and Little Bear), is referred to in Job xxvi. 13: "By His Spirit He hath garnished the heavens; His hand hath formed the crooked serpent (R.V. swift. Marg. *fleeing* or *gliding*. See Is. xxvii. 1; xliii. 14)." This word "garnished" is peculiar. The R.V. puts in the margin, *beauty*. In Ps. xvi. 6, it is rendered *goodly*. "I have a goodly heritage." In Dan. iv. 2, it is rendered, "I thought

* Note the structure of this verse:—

A | The seven stars,
B | Orion,
A | The twelve signs,
B | Arcturus.

it good to show," referring to "the signs and wonders" with which God had visited Nebuchadnezzar. It appears from this that God "*thought it good to show*" by these signs written in the heavens the wonders of His purposes and counsels, and it was by His Spirit that He made it known; it was His hand that *coiled* (חולל) the crooked serpent among the stars of heaven.

Thus we see that the Scriptures are not silent as to the great antiquity of the signs and constellations.

If we turn to history and tradition, we are at once met with the fact that the Twelve Signs are the same, both as to the meaning of their names and as to their order *in all the ancient nations of the world.* The Chinese, Chaldean, and Egyptian records go back to more than 2,000 years B.C. Indeed, the Zodiacs in the Temples of Denderah and Esnéh, in Egypt, are doubtless copies of Zodiacs still more ancient, which, from internal evidence, must be placed nearly 4,000 B.C., when the summer solstice was in Leo.

Josephus hands down to us what he gives as the traditions of his own nation, corroborated by his reference to eight ancient Gentile authorities, whose works are lost. He says that they all assert that "God gave the antediluvians such long life that they might perfect those things which they had invented in astronomy." Cassini commences his *History of Astronomy* by saying "It is impossible to doubt that astronomy was invented from the beginning of the world; history, profane as well as sacred, testifies to this truth." Nouet, a French astronomer, infers that the Egyptian Astronomy must have arisen 5,400 B.C.!

Ancient Persian and Arabian traditions ascribe its invention to Adam, Seth, and Enoch. Josephus asserts that it originated in the family of Seth; and he says that the children of Seth, and especially Adam, Seth, and Enoch, that their revelation might not be lost as to the two coming judgments of Water and Fire, made two pillars (one of brick, the other of stone), describing the whole of the predictions of the stars upon them, and in case the brick pillar should be destroyed by the flood, the stone would preserve the revelation (Book i. chs. 1–3).

This is what is doubtless meant by Gen. xi. 4," And they said, Go to, let us build us a city and a tower whose top *may reach* unto heaven." The words "*may reach*" are in italics. There is nothing in the verse which relates to the height of this tower. It merely says וְרֹאשׁוֹ בַשָּׁמַיִם, *and his top with the heavens*, *i.e.* with the pictures and the stars, just as we find them in the ancient temples of Denderah and Esnéh in Egypt. This tower, with its planisphere and pictures of the signs and constellations, was to be erected like those temples were afterwards, in order to preserve the revelation, "lest we be scattered abroad upon the face of the whole earth."

This is corroborated by Lieut.-Gen. Chesney, well known for his learned researches and excavations among the ruins of Babylon, who, after describing his various discoveries, says,* "About five miles S.W. of Hillah, the most remarkable of all the ruins, the *Birs Nimroud* of

* General Chesney allowed the late Dean Goode to copy the passage, among other matters, from his private MS. The Dean quotes it in his *Warburtonian Lectures* (2nd Ed., Note I. to Sermon IV., p. 170–1).

the Arabs, rises to a height of 153 feet above the plain from a base covering a square of 400 feet, or almost four acres. It was constructed of kiln-dried bricks in seven stages to correspond with the planets to which they were dedicated: the lowermost black, the colour of Saturn; the next orange, for Jupiter; the third red, for Mars; and so on.* These stages were surmounted by a lofty tower, on the summit of which, we are told, were the signs of the Zodiac and other astronomical figures; thus having (as it should have been translated) *a representation of the heavens*, instead of 'a top which reached unto heaven.'"

This Biblical evidence carries us at once right back to the Flood, or about 2,500 years B.C.

This tower or temple, or both, was also called "*The Seven Spheres*," according to some; and "The Seven Lights," according to others. It is thus clear that the popular idea of its height and purpose must be abandoned, and its astronomical reference to revelation must be admitted. The tower was an attempt to preserve and hand down the antediluvian traditions; their sin was in keeping together instead of scattering themselves over the earth.

Another important statement is made by Dr. Budge, of the British Museum.† He says, "It must never be forgotten that the Babylonians were a nation of star-gazers, and that they kept a body of men to do nothing else but report eclipses, appearances of the moon, sun-spots, etc., etc."

---

* Fragments of these coloured glazed bricks are to be seen in the British Museum.

† *Babylonian Life and History*, p. 36.

"Astronomy, mixed with astrology, occupied a large number of tablets in the Babylonian libraries, and Isaiah, xlvii. 13, refers to this when he says to Babylon, 'Thou art wearied in the multitude of thy counsels. Let now thy astrologers (marg. *viewers of the heavens*), the star-gazers, the monthly prognosticators stand up.' The largest astrological work of the Babylonians contained seventy tablets, and was compiled by the command of Sargon of Agade thirty-eight hundred years before Christ! It was called the 'Illumination of Bel.'"

"Their observations were made in towers called "ziggurats" (p. 106).

"They built observatories in all the great cities, and reports like the above [which Dr. Budge gives in full] were regularly sent to the King" (p. 110).

"They were able to calculate eclipses, and had long lists of them." "They found out that the sun was spotted, and they knew of comets." "They were the inventors of the Zodiac" (?). There are fragments of two (ancient Babylonian) planispheres in the British Museum with figures and calculations inscribed upon them. "The months were called after the signs of the Zodiac" (p. 109).

We may form some idea of what this "representation of the heavens" was from the fifth "Creation Tablet," now in the British Museum. It reads as follows:—

"Anu [*the Creator*] made excellent the mansions [i.e. *the celestial houses*] of the great gods [twelve] in number [i.e. *the twelve signs or mansions of the sun*].

The stars he placed in them. The lumasi [i.e. *groups of stars or figures*] he fixed.

He arranged the year according to the bounds [i.e. *the twelve signs*] which he defined.
For each of the twelve months three rows of stars [i.e. *constellations*] he fixed.
From the day when the year issues forth unto the close, he marked the mansions [i.e. *the Zodiacal Signs*] of the wandering stars [i.e. *planets*] to know their courses that they might not err or deflect at all."

Coming down to less ancient records: EUDOXOS, an astronomer of Cnidus (403 to 350 B.C.), wrote a work on Astronomy which he called *Phainomena.* ANTIGONUS GONATAS, King of Macedonia (273–239 B.C.), requested the Poet ARATUS to put the work of EUDOXUS into the form of a poem, which he did about the year 270 B.C. ARATUS called his work *Diosemeia* (*the Divine Signs*). He was a native of Tarsus, and it is interesting for us to note that his poem was known to, and, indeed, must have been read by, the Apostle Paul, for he quotes it in his address at Athens on Mars' Hill. He says (Acts xvii. 28), "For in Him we live, and move, and have our being; as certain also of your own poets have said, For we are also his offspring."* Several translations of this poem have been made, both by CICERO and others, into Latin, and in recent times into English by E. Poste, J. Lamb, and others. The following is the opening from the translation of Robert Brown, jun.:—

"From Zeus we lead the strain; he whom mankind
Ne'er leave unhymned: of Zeus all public ways,
All haunts of men, are full; and full the sea,
And harbours; and of Zeus all stand in need.
We are his offspring:* and he, ever good and mild to man,

* τοῦ γαρ καὶ γένος ἐσμεν.

Gives favouring signs, and rouses us to toil.
Calling to mind life's wants: when clods are best
For plough and mattock: when the time is ripe
For planting vines and sowing seeds, he tells,
Since he himself hath fixed in heaven these Signs,
The stars dividing: and throughout the year
Stars he provides to indicate to man
The seasons' course, that all things duly grow," etc., etc.

Then ARATUS proceeds to describe and explain all the Signs and Constellations as the Greeks in his day understood, or rather misunderstood, them, after their true meaning and testimony had been forgotten.

Moreover, ARATUS describes them, not as they were seen in his day, but as they were seen some 4,000 years before. The stars were not seen from Tarsus as he describes them, and he must therefore have written from a then ancient Zodiac. For notwithstanding that we speak of "fixed stars," there is a constant, though slow, change taking place amongst them. There is also another change taking place owing to the slow recession of the pole of the heavens (about 50″ in the year); so that while *Alpha* in the constellation of *Draco* was the Polar Star when the Zodiac was first formed, the Polar Star is now *Alpha* in what is called *Ursa Minor*. This change alone carries us back at least 5,000 years. The same movement which has changed the relative position of these two stars has also caused the constellation of the *Southern Cross* to become invisible in northern latitudes. When the constellations were formed the *Southern Cross* was visible in N. latitude 40°, and was included in their number. But, though known by tradition, it had not been seen in that latitude for some twenty centuries, until the Cape of Good Hope had been discovered. Then was seen again *The Southern Cross* depicted by the Patriarchs. Here is another indisputable proof as to the antiquity of the formation of the Zodiac.

PTOLEMY (150 A.D.) transmits them from HIPPARCHUS (130 B.C.) "as of unquestioned authority, unknown origin, and unsearchable antiquity."

Sir William Drummond says that "the traditions of the Chaldean Astronomy seem the fragments of a mighty system fallen into ruins."

The word *Zodiac* itself is from the Greek Ζωδιακός, which is not from ζάω, *to live*, but from a primitive root through the Hebrew *Sodi*, which in Sanscrit means *a way*. Its etymology has no connection with *living creatures*, but denotes *a way*, or *step*, and is used of the *way* or *path* in which the sun appears to move amongst the stars in the course of the year.

To an observer on the earth the whole firmament, together with the sun, appears to revolve in a circle once in twenty-four hours. But the time occupied by the stars in going round, differs from the time occupied by the sun. This difference amounts to about one-twelfth part of the whole circle in each month, so that when the circle of the heavens is divided up into twelve parts, the sun appears to move each month through one of them. This path which the sun thus makes amongst the stars is called the *Ecliptic*.*

---

* Besides this *monthly* difference, there is an *annual* difference; for at the end of twelve months the sun does not come back to exactly the same point in the sign which commenced the year, but is a little behind it. But this difference, though it

Each of these twelve parts (consisting each of about 30 degrees) is distinguished, not by numbers or by letters, but by pictures and names, and this, as we have seen, from the very earliest times. They are preserved to the present day in our almanacs, and we are taught their order in the familiar rhymes:—

"The RAM, the BULL, the heavenly TWINS,
And next the CRAB, the LION shines,
The VIRGIN and the SCALES;
The SCORPION, ARCHER, and SEA-GOAT,
The MAN that carries the Water-pot,
And FISH with glittering scales."

These signs have always and everywhere been preserved in this order, and have begun with ARIES. They have been known amongst all nations, and in all ages, thus proving their common origin from one source.

---

occurs every year, is so small that it will take 25,579 years for the sun to complete this vast cycle, which is called *The precession of the Equinoxes; i.e.*, about one degree in every 71 years. If the sun came back to the precise point at which it began the year, each *sign* would correspond, always and regularly, exactly with a particular *month;* but, owing to this constant regression, the sun (while it goes through the whole twelve signs every year) commences the year in one sign for only about 2,131 years. In point of fact, since the Creation the commencement of the year has changed to the extent of nearly three of the signs. When Virgil sings—

"*The White Bull with golden horns opens the year,*"

he does not record what took place in his own day. This is another proof of the antiquity of these signs.

The *Ecliptic*, or path of the sun, if it could be viewed from immediately beneath the Polar Star, would form a complete and perfect circle, would be concentric with the *Equator*, and all the stars and the sun would appear to move in this circle, never rising or setting. To a person north or south of the Equator the stars therefore rise and set obliquely; while to a person on the Equator they rise and set perpendicularly, each star being twelve hours above and twelve below the horizon.

The points where the two circles (the *Ecliptic* and the *Equator*) intersect each other are called the *Equinoctial points*. It is the movement of these points (which are now moving from Aries to Pisces) which gives rise to the term, "*the precession of the Equinoxes.*"

The figures themselves are perfectly arbitrary. There is nothing in the groups of stars to even suggest the figures. This is the first thing which is noticed by every one who looks at the constellations. Take for example the sign of VIRGO, and look at the stars. There is nothing whatever to suggest a human form; still less is there anything to show whether that form is a man or a woman. And so with all the others.

The *picture*, therefore, is the original, and must have been drawn around or connected with certain stars, simply in order that it might be identified and associated with them; and that it might thus be remembered and handed down to posterity.

There can be no doubt, as the learned Authoress of *Mazzaroth* conclusively proves, that these signs were afterwards identified with the twelve sons of Jacob. Joseph sees the sun and moon and eleven stars bowing down to him, he himself being the twelfth (Gen. xxxvii. 9). The blessing of Jacob (Gen. xlix.) and the blessing of Moses (Deut. xxxiii.) both bear witness to the existence of these signs in their day. And it is more than probable that each of the Twelve Tribes bore one of them on its standard. We read in Num. ii. 2, "Every man of the children of Israel shall pitch by his own STANDARD, with the ENSIGN of their father's house" (R.V. "with the ensigns of their fathers' houses"). This "Standard" was the *Degel* (דֶּגֶל) on which the "Sign" (אוֹת, *Oth*) was depicted. Hence it was called the "*En-sign.*" Ancient Jewish authorities declare that each tribe had one of the signs as its own, and it is highly probable, even from Scripture, that four of the tribes carried its "Sign"; and that these four were placed at the four sides of the camp.

If the Lion were appropriated to Judah, then the other three would be thus fixed, and would be the same four that equally divide the Zodiac at its four cardinal points. According to Num. ii. the camp was thus formed:—

EAST.

ISSACHAR (*Cancer*). — JUDAH The Lion (*Leo*). — ZEBULON (*Virgo*).

NORTH. NEPHTALI (*Capricornus*). — DAN The Scorpion (*Scorpio*). — ASHER (*Sagittarius*).

LEVI The Scales (*Libra*).

SOUTH. SIMEON (*Pisces*). — REUBEN The Man (*Aquarius*). — GAD (*Aries*).

BENJAMIN (*Gemini*). — EPHRAIM and MANASSEH The Bull (the two horns of *Taurus*).

WEST.

If the reader compares the above with the blessings of Israel and Moses, and compares the meanings and descriptions given below with those blessings, the connection will be clearly seen. Levi, for example, had no standard, and he needed none, for he kept "the balance of the Sanctuary," and had the charge of that brazen altar on which the atoning blood outweighed the nation's sins.

The four great signs which thus marked the four sides of the camp, and the four quarters of the Zodiac, are the same four which form the Cherubim (the Eagle, the Scorpion's enemy, being substituted for the Scorpion). The Cherubim thus form a compendious expression of the hope of Creation, which, from the very first, has been bound up with the Coming One, who alone should cause its groanings to cease.

But this brings us to the Signs themselves and their interpretation.

These pictures were designed to preserve, expound, and perpetuate the one first great promise and prophecy of Gen. iii. 15, that all hope for Man, all hope for Creation, was bound up in *a coming Redeemer;* One who should be born of a woman; who should first suffer, and afterwards gloriously triumph; One who should first be wounded by that great enemy who was the cause of all sin and sorrow and death, but who should finally crush the head of "that Old Serpent the Devil."

These ancient star-pictures reveal this Coming One. They set forth "the sufferings of Christ and the glory that should follow." Altogether there are forty-eight of them, made up of twelve SIGNS, each sign containing three CONSTELLATIONS.

These may be divided into *three* great books, each

B 2

book containing four chapters (or Signs); and each chapter containing three sections (or Constellations).

Each book (like the four Gospels) sets forth its peculiar aspect of the Coming One; beginning with the promise of His coming, and ending with the destruction of the enemy.

But where are we to *begin* to read this wondrous Heavenly Scroll? A circle has proverbially neither beginning nor end. In what order then are we to consider these signs? In the heavens they form a never-ending circle. Where is the beginning and where is the end of this circle through which the sun is constantly moving? Where are we to break into this circle? and say, *This is the commencement*. It is clear that unless we can determine this original starting point we can never read this wondrous book aright.

As I have said, the popular beginning to-day is with ARIES, *the* Ram. But comparing this Revelation with that which was afterwards written "in the Volume of the Book," VIRGO is the only point where we can intelligently begin, and LEO is the only point where we can logically conclude. Is not this what is spoken of as the unknown and insoluble mystery—"The riddle of the SPHINX"? The word "Sphinx" is from σφίγγω, *to bind closely together*. It was therefore designed to show where the two ends of the Zodiac were to be joined together, and where the great circle of the heavens begins and ends.

The SPHINX is a figure with the *head of a woman* and the *body of a lion!* What is this but a never-ceasing monitor, telling us to begin with *Virgo* and

"LEO." "THE SPHINX." "VIRGO."

The Signs of LEO and VIRGO, from the ceiling of the Portico of the Temple of ESNEH, showing the SPHINX between, uniting the beginning and end of the Zodiac.

to end with *Leo!* In the Zodiac in the Temple of Esnéh, in Egypt, a Sphinx is actually placed between the Signs of Virgo and Leo, as shown in the illustration on the preceding page. It is a tracing from the drawing of Signor Bossi, executed on the spot, under the direction of the late Mr. Edward J. Cooper, in 1820.

Beginning, then, with VIRGO, let us now spread out the contents of this Heavenly Volume, so that the eye can take them in at a glance. Of course we are greatly hindered in this, in having to use the modern Latin names which the Constellations bear to-day.* Some of these names are mistakes, others are gross perversions of the truth, as proved by the pictures themselves, which are far more ancient, and have come down to us from primitive times.

After the Revelation came to be written down in the Scriptures, there was not the same need for the preservation of the Heavenly Volume. And after the nations had lost the original meaning of the pictures, they invented a meaning out of the vain imagination of the thoughts of their hearts. The Greek Mythology is an interpretation of (only some of) the signs and constellations after their true meaning had been forgotten. It is popularly believed that Bible truth is an evolution from, or development of, the ancient religions of the world. But the fact is that they themselves are a *corruption* and *perversion of primitive truth!*

---

* It is exactly the same with the books of the Bible. Their order and their names, *as we have them* in the English Bible, are those which *man* has given them, copied from the Septuagiut and Vulgate, and in many cases are not the Divine names according to the Hebrew Canon.

We will now give *the contents* of this Heavenly Volume of Divine Revelation, and afterwards proceed to develope it, explain it in detail, and compare it with the same truth which was afterwards written down in the Scriptures.

The First Book.

THE REDEEMER.

(His First Coming.)

"The Sufferings of Christ."

CHAPTER I.

The Prophecy of the Promised Seed of the Woman.

VIRGO (*The Virgin. A woman bearing a branch in her right hand and an ear of corn in her left*). The Promised Seed of the woman.

§ 1. Coma (*The Desired. The woman and child*). The Desired of all nations.

§ 2. Centaurus (*The Centaur with two natures, holding a spear piercing a victim*). The despised sin offering.

§ 3. Boötes (*a man walking bearing a branch called* ARCTURUS, *meaning the same*). He cometh.

CHAPTER II.

The Redeemer's Atoning Work.

LIBRA (*The Scales*). The price deficient balanced by the price which covers.

§ 1. Crux, *The Cross* endured.

§ 2. Lupus, or Victima, *The Victim* slain.

§ 3. Corona, *The Crown* bestowed.

CHAPTER III.

THE REDEEMER'S CONFLICT.

SCORPIO (*The Scorpion*) seeking to wound, but itself trodden under foot.

§ 1. SERPENS (*The Serpent* struggling with the man).

§ 2. O-PHI-U-CHUS (*The man* grasping the serpent). The struggle with the enemy.

§ 3. HERCULES (*The mighty man. A man kneeling on one knee, humbled in the conflict, but holding aloft the tokens of victory, with his foot on the head of the Dragon*). The mighty Vanquisher seeming to sink in the conflict.

CHAPTER IV.

THE REDEEMER'S TRIUMPH.

SAGITTARIUS (*The Archer*). The Two-natured Conqueror going forth "Conquering and to conquer."

§ 1. LYRA (*The Harp*). Praise prepared for the Conqueror.

§ 2. ARA (*The Altar*). Consuming fire prepared for His enemies.

§ 3. DRACO (*The Dragon*). The Old Serpent—the Devil, cast down from heaven.

The Second Book.

THE REDEEMED.

THE RESULT OF THE REDEEMER'S SUFFERINGS.

CHAPTER I.

THEIR BLESSINGS PROCURED.

CAPRICORNUS (*The fish-goat*). The goat of Atonement slain for the Redeemed.

§ 1. Sagitta (*The Arrow*). The arrow of God sent forth.

§ 2. Aquila (*The Eagle*). The smitten One falling.

§ 3. Delphinus (*The Dolphin*). The dead One rising again.

CHAPTER II.

Their Blessings Ensured.

AQUARIUS (*The Water-Bearer*). The living waters of blessing poured forth for the Redeemed.

§ 1. Piscis Australis (*The Southern Fish*). The blessings bestowed.

§ 2. Pegasus (*The Winged Horse*). The blessings quickly coming.

§ 3. Cygnus (*The Swan*). The Blesser surely returning

CHAPTER III.

Their Blessings in Abeyance.

PISCES (*The Fishes*). The Redeemed blessed though bound.

§ 1. The Band—bound, but binding their great enemy Cetus, the sea monster.

§ 2. Andromeda (*The Chained Woman*). The Redeemed in their bondage and affliction.

§ 3. Cepheus (*The King*). Their Redeemer coming to rule.

CHAPTER IV.

Their Blessings Consummated and Enjoyed.

ARIES (*The Ram or Lamb*). The Lamb that was slain, prepared for the victory.

§ 1. Cassiopeia (*The Enthroned Woman*). The captive delivered, and preparing for her husband, the Redeemer.

§ 2. Cetus (*The Sea Monster*). The great enemy bound.

§ 3. Perseus (*The Breaker*). Delivering His redeemed.

The Third Book.

THE REDEEMER.

(His Second Coming.)

"The glory that should follow."

CHAPTER I.

Messiah, the coming Judge of all the Earth.

TAURUS (*The Bull*). Messiah coming to rule.

§ 1. Orion, *Light breaking forth* in the person of the Redeemer.

§ 2. Eridanus (*The River of the Judge*). Wrath breaking forth for His enemies.

§ 3. Auriga (*The Shepherd*). Safety for the Redeemed in the day of that wrath.

CHAPTER II.

Messiah's Reign as Prince of Peace.

GEMINI (The Twins). The twofold nature of the King.

§ 1. Lepus (*The Hare*), or the enemy trodden under foot.

§ 2. Canis major (*The Dog*), or Sirius, the coming glorious Prince of Princes.

§ 3. Canis Minor (*The Second Dog*), or Procyon, the exalted Redeemer.

CHAPTER III.

Messiah's Redeemed Possessions.

CANCER (The Crab). The possessions held fast.

§ 1. Ursa Minor (*The Lesser Bear*). The lesser sheepfold.

§ 2. Ursa Major (*The Great Bear*). The fold and the flock.

§ 3. Argo (*The Ship*). The redeemed pilgrims safe at home.

CHAPTER IV.

Messiah's Consummated Triumph.

LEO (*The Lion*). The Lion of the Tribe of Judah aroused for the rending of the Enemy.

§ 1. Hydra (*The Serpent*). That old Serpent—the Devil, destroyed.

§ 2. Crater (*The Cup*). The cup of Divine wrath poured out upon him.

§ 3. Corvus (*The Crow, or Raven*). Birds of prey devouring him.

Such are the contents of this wondrous book that is written in the heavens. Thus has God been speaking and emphasizing and developing His first great prophetic promise of Gen. iii. 15.

Though for more than 2,500 years His people had not this Revelation written in a book as we now have it in the Bible, they were not left in ignorance and darkness as to God's purposes and counsels; nor were they without hope as to ultimate deliverance from all evil and from the Evil One.

Adam, who first heard that wondrous promise, repeated it, and gave it to his posterity as a most precious heritage—the ground of all their faith, the substance of all their hope, the object of all their desire. Seth and Enoch took it up. Enoch, we know, prophesied of the Lord's coming, saying, "Behold the Lord cometh with ten thousands of His saints to execute judgment upon all" (Jude 14). How could these "holy prophets, since the world began," have recorded their prophecies better, or more effectually, or more truthfully and powerfully, than in these star-pictures and their interpretation? This becomes a certainty when we remember the words of the Holy Spirit by Zacharias (Luke i. 67–70):—

> "Blessed be the Lord God of Israel;
> For He hath visited and redeemed His people,
> And hath raised up a horn of salvation for us
> In the house of His servant David;
> As He spake by the mouth of HIS HOLY PROPHETS
> WHICH HAVE BEEN SINCE THE WORLD BEGAN."

The same truth is revealed through Peter, in Acts iii. 20, 21:—"He shall send Jesus Christ, which before was preached unto you; whom the heaven must receive until the times of restitution of all things, which God hath spoken by the mouth of all HIS HOLY PROPHETS SINCE THE WORLD BEGAN."

These words have new meaning for us, if we see the things which were spoken "since the world began," thus written in the heavens, which utter speech (*i.e.* prophecy), and show forth this knowledge day after day and night after night, the heritage of all the earth, and their words reaching unto the ends of the world.

This Revelation, coinciding as it does in all its facts and truths with that afterwards recorded "in the Volume of the Book," must have had the same Divine origin, must have been made known by the inspiration of the same Holy Spirit.

We now proceed to compare the two, and we shall see how they agree at every point, proving that the source and origin of this Divine Revelation is one and the same.

I. VIRGO. (the Virgin)
ECLIPTIC
LEO
CRATER
Corvus
Denebola
Al Mureddin or Vindemiatrix
Al Zimach
LIBRA
Boötes

The First Book.

THE REDEEMER.

(His First Coming.)

"The Sufferings of Christ."

THE First Book is occupied with the PERSON of the Coming One. It covers the whole ground, and includes the conflict and the victory of the Promised Seed, but with special emphasis on His Coming. The book opens with the promise of His coming, and it closes with the Dragon cast down from heaven.

CHAPTER I.

The sign VIRGO.

The Promised Seed of the Woman.

Here is the commencement of all prophecy in Gen. iii. 15, spoken to the serpent:—"I will put enmity between thee and the woman, and between thy seed and her seed: it shall bruise thy head, and thou shalt bruise His heel." This is the prophetic announcement which the Revelation in the heavens and in the Book is designed to unfold and develope. It lies at the root of all the ancient traditions and mythologies, which are simply the perversion and corruption of primitive truth.

Virgo is represented as a woman with a *branch* in her right hand, and some ears of corn in her left hand. Thus giving a two-fold testimony of the Coming One.

The name of this sign in the Hebrew is *Bethulah*, which means *a virgin*, and in the Arabic *a branch*. The two words are connected, as in Latin—*Virgo*, which means *a virgin;* and *virga*, which means *a branch* (Vulg. Isa. xi. 1). Another name is *Sunbul*, Arabic, *an ear of corn*.

In Gen. iii. 15 she is presented only as a woman; but in later prophecies her nationality is defined as being of the stock of Israel, the seed of Abraham, the line of David; and, further, she is to be a virgin. There are two prominent prophecies of her and her seed: one is connected with the first coming in incarnation, Isa. vii. 14 (quoted in Matt. i. 23).

> "Behold, a virgin shall conceive and bear a son,
> And shall call his name Immanuel."

The other is connected with His second coming, leaping over the sufferings and this present interval of His rejection, and looking forward to His coming in glory and judgment, Isa. ix. 6, 7 (quoted in Luke ii. 11 and i. 32, 33)—

> "For unto us a child is born,
> Unto us a son is given;*
> And the government shall be upon His shoulder;
> And His name shall be called Wonderful, Counsellor,
> The Mighty God, The Everlasting Father, The Prince of Peace.
> Of the increase of His government there shall be no end.

---

* Here, the fact of His humiliation, together with this long period of His rejection, is leaped over, and the prophecy passes on at once—over at least a period of 1893 years—to this "glory which should follow."

Upon the throne of David, and upon His kingdom,
To order it, and to establish it
With judgment and with justice
From henceforth even for ever.
The zeal of the LORD of hosts will perform this."

It is difficult to separate the Virgin and her Seed in the prophecy; and so, here, we have first the sign VIRGO, where the name points to her as the prominent subject; while in the first of the three constellations of this sign, where the woman appears again, the name COMA points to the child as the great subject.

Virgo contains 110 stars, *viz.*, one of the 1st magnitude, six of the 3rd, ten of the 4th, etc.

ARATUS thus sings of them:—

"Beneath Boötes feet the Virgin seek,
Who carries in her hand a glittering spike. . . .
Over her shoulder there revolves a star
In the right wing, superlatively bright;*
It rolls beneath the tail, and may compare
With the bright stars that deck the Greater Bear.
Upon her shoulder one bright star is borne,†
One clasps the circling girdle of her loins,‡
One at her bending knee; § and in her hand
Glitters that bright and golden Ear of Corn.||

Thus the brightest star in VIRGO (α)¶ has an ancient name, handed down to us in all the star-maps, in which the Hebrew word (צֶמַח) *Tsemech* is preserved. It is

---

* ε, *Al Mureddin.* † β, *Zavijavah.* ‡ The star now marked δ.
§ The star ζ. || The star α, *Al Zimach.*

¶ The stars are known by Greek letters and sometimes by numbers, &c. Alpha (α) denotes a star of the *first* magnitude; Beta (β), the second, and so on. This plan was originated by Bayer in his *Uranometria*, 1603. The star *Alpha*, as seen in the New Great Equatorial Telescope recently set up at Greenwich, is now discovered to be really *a double* star, though it had hitherto always appeared to be *one*.

called in Arabic *Al Zimach*, which means *the branch*. This star is in the ear of corn which she holds in her left hand. Hence the star has a modern Latin name, which has almost superseded the ancient one, *Spica*, which means, *an ear of corn*. But this hides the great truth revealed by its name *Al Zimach*. It foretold the coming of Him who should bear this name. The same Divine inspiration has, in the written Word, four times connected it with Him. There are twenty Hebrew words translated "Branch," but only one of them (*Tsemech*) is used exclusively of the Messiah, and this word only four times.* Each of these further connects Him with one special account of Him, given in the Gospels.

(1.) Jer. xxiii. 5, 6—

> "Behold, the days come, saith the LORD,
> That I will raise unto David a righteous BRANCH
> (*i.e.*, a Son),
> And a KING shall reign and prosper."

The account of His coming as King is written in the Gospel according to Matthew, where Jehovah says to Israel, "Behold thy KING." (Zech. ix. 9; Matt. xxi. 9.)

(2.) Zech. iii. 8.—"Behold I will bring forth my SERVANT the BRANCH." In the Gospel according to Mark we find the record of Jehovah's servant and His service, and we hear Jehovah's voice saying, "Behold my SERVANT." (Isa. xlii. 1.)

(3.) Zech. vi. 12.—"Thus speaketh the LORD of hosts, saying, Behold the MAN whose name is the

* Jer. xxxiii. 15 being only a repetition of Jer. xxiii. 5.

BRANCH." In the Gospel according to Luke we behold Him, presented in "the MAN Christ Jesus."

4. Isa. iv. 2.—"In that day shall the BRANCH of JEHOVAH be beautiful and glorious." So that this Branch, this Son, is Jehovah Himself; and as we read the record of John we hear the voice from heaven saying, "Behold your GOD." (Isa. xl. 9.)

This is the Branch foretold by the star *Al Zimach* in the ear of corn.

The star β is called *Zavijaveh*, which means *the gloriously beautiful*, as in Isa. iv. 2. The star ϵ, in the arm bearing the branch, is called *Al Mureddin*, which means *who shall come down* (as in Ps. lxxii. 8), or *who shall have dominion.* It is also known as *Vindemiatrix*, a Chaldee word which means *the son*, or *branch, who cometh.*

Other names of stars in the sign, not identified, are—

Subilah, who carries. (Isa. xlvi. 4.)
Al Azal, the Branch. (As in Isa. xviii. 5.)
Subilon, a spike of corn. (As in Isa. xvii. 5.)

The Greeks, ignorant of the Divine origin and teaching of the sign, represented Virgo as *Ceres*, with ears of corn in her hand.

In the Zodiac in the Temple of Denderah, in Egypt, about 2000 B.C. (now in Paris), she is likewise represented with a branch in her hand, but ignorantly explained by a false religion to represent *Isis!* Her name is called *Aspolia*, which means *ears of corn*, or *the seed*, which shows that though the woman is seen, it is her Seed who is the great subject of the prophecy.

Passing to the three constellations anciently assigned to the sign Virgo, we come to what may be compared to *three sections* of the chapter, each giving some further detail as to the interpretation of its teaching.

§ 1. COMA (The Woman and Child).

The Desired of all Nations.

The first constellation in Virgo explains that this coming "Branch" will be a child, and that He should be the "Desire of all nations."

The ancient name of this constellation is *Comah*,* *the desired*, or *the longed for*. We have the word used by the Holy Spirit in this very connection, in Hag. ii. 7: "The DESIRE of all nations shall come."

The ancient Zodiacs pictured this constellation as a woman with a child in her arms. ALBUMAZAR† (or ABU MASHER), an Arabian astronomer of the eighth century, says, "There arises in the first Decan,‡ as the Persians, Chaldeans, and Egyptians, and the two HERMES and ASCALIUS teach, *a young woman*, whose Persian name denotes a pure virgin, sitting on a throne, *nourishing an infant boy* (the boy, I say), having a Hebrew name, by some nations called

* From כָּמַהּ, which occurs only in Ps. lxiii. 1, "my flesh *longeth* for thee." It is akin to חָמַד, *to desire*. Ps. xix. 10; Is. liii. 2; Hag. ii. 7; etc.

† A Latin translation of his work is in the British Museum Library. He says the Persians understood these signs, but that the Indians perverted them with inventions.

‡ The constellations are called *Decans*. The word means *a part*, and is used of the three parts into which each sign is divided, each of which is occupied by a constellation.

URSA MAJOR
Denebola
LEO
II. COMA. (the Desired)
BOÖTES
VIRGO

IHESU, with the signification IEZA, which in Greek is called CHRISTOS."

But this picture is not found in any of the *modern* maps of the stars. There we find to-day a woman's wig! It appears that BERENICE, the wife of EUERGETES (PTOLEMY III.), king of Egypt in the third century B.C., when her husband once went on a dangerous expedition, vowed to consecrate her fine head of hair to Venus if he returned in safety. Her hair, which was hung up in the Temple of Venus, was subsequently stolen, and to comfort BERENICE, CONON, an astronomer of Alexandria (B.C. 283–222), gave it out that Jupiter had taken it and made it a constellation!

This is a good example of how the meaning of other constellations have been perverted (ignorantly or intentionally). In this case, as in others, the transition from ancient to more modern languages helped to hide the meaning. The Hebrew name was COMA (*desired*). But the Greeks had a word for hair, *Có-me.* This again is transferred to the Latin *coma*, and thus "*Coma Berenicæ*" (*the hair of Berenice*) comes down to us to-day as the name of this constellation, and gives us a woman's wig instead of that Blessed One, "the Desire of all Nations."

In this case, however, we are able to give absolute proof that this is a perversion.

The ancient Egyptian name for this constellation was *Shes-nu*, *the desired son!*

The Zodiac in the Temple of Denderah, in Egypt, going back at least 2,000 years B.C., has no trace of any hair, but it has the figure of a woman and child.

C 2

In our illustration we have given a copy of this very ancient picture, and not the wig of hair!

We have been permitted to trace it from a work on *Egyptian Scenery* by the late eminent astronomer, Edward J. Cooper, of Markree Castle, co. Sligo, who visited that Temple in the year 1820 with an Italian artist, Signor Bossi. The original drawing from which our tracing is made (and enlarged) was drawn by Signor Bossi on the spot, before it was taken to Paris in 1821.* We thus have before us the exact representations of one of these star-pictures at least 4,000 years old.

Even Shakespeare understood the truth about this constellation picture, which has been so long covered by modern inventions. In his *Titus Andronicus*† he speaks of an arrow being shot up to heaven to the "*Good boy in Virgo's lap.*"

The constellation itself is very remarkable. Others contain one or two stars of the first or second magnitude, and then a greater or less variety of lesser stars; but this is peculiar from having no one very bright star, but contains so many stars of the 4th and 5th magnitudes. It contains 43 stars altogether, ten being of the 4th magnitude, and the remainder of the 5th, 6th, etc.

It was in all probability the constellation of *Coma* in which "the Star of Bethlehem" appeared. There was a traditional prophecy, well-known in the East,

---

* It appears that MM. Saulnier, fils, and Lelorrain arrived while Signor Bossi was engaged in copying it, but concealed their design to remove it. The King of France paid £6,250 sterling for it. It has since been copied, and lithographs have been published.

† Act IV., Scene 3.

carefully preserved and handed down, that a new star would appear in this sign when He whom it foretold should be born.

This was, doubtless, referred to in the prophecy of Balaam, which would thus receive a double fulfilment, first of the literal "Star," and also of the person to whom it referred. The Lord said by Balaam (Num. xxiv. 17),

> "There shall come* a star out of Jacob,
> And a sceptre shall rise out of Israel."

Thomas Hyde, an eminent Orientalist (1636–1703), writing on the ancient religion of the Persians, quotes from ABULFARAGIUS (an Arab Christian Historian, 1226–1286), who says that ZOROASTER, or ZERDUSHT, the Persian, was a pupil of Daniel the Prophet, and that he predicted to the Magians (who were the astronomers of Persia), that when they should see *a new star* appear it would notify the birth of a mysterious child, whom they were to adore. It is further stated in the *Zend Avesta* that this new star was to appear in the sign of the Virgin. Some have supposed that this passage is not genuine. But whether it was interpolated before or after the event, it is equally good evidence for our purpose here. For if it was written *before* the event, it is evidence of the *prophetic announcement;* and if it was interpolated *after* the event it is evidence of the *historic fact.*

The Book of Job shows us how Astronomy flourished

---

* *I.e., come forth* (as in the R.V.). *At,* as the preposition מ is rendered in Gen. iii. 24. "There shall come forth a star at or over the inheritance or possessions of Jacob," thus indicating the locality which would be on the *meridian* of this star.

in Idumea; and the Gospel according to Matthew shows that the Persian Magi, as well as others, were looking for "the Desire of all nations."

New stars have appeared again and again. It was in 125 B.C. that a star, so bright as to be seen in the day-time, suddenly appeared. It was this that caused HIPPARCHUS to draw up his catalogue of stars, which has been handed down to us by PTOLEMY (150 A.D.).

This new star would show the *latitude*, passing at that time immediately overhead at midnight, every twenty-four hours; while the prophecy would give the *longitude* as the land of Jacob. Having these two factors, it would be only a matter of observation, and easy for the Magi to find the place where it would be vertical, and thus to locate the very spot of the birth of Him of whom it was the sign, for they emphatically called it "His Star." There is a beautiful tradition which relates how, in their difficulty, on their way from Jerusalem to find the actual spot under the *Zenith* of this star, these Magi sat down beside David's "Well of Bethlehem" to refresh themselves. There they saw the star reflected in the clear water of the well. Hence it is written that "when they saw the star they rejoiced with exceeding joy," for they knew they were at the very spot and place of His appearing whence He was to "come forth."

There can be little doubt that it was *a new star*. In the first place a new star is no unusual phenomenon. In the second place the tradition is well supported by ancient Christian writers. One speaks of its "surpassing brightness." Another (IGNATIUS, Bp. of Antioch, A.D. 69) says, "At the appearance of the Lord a star shone forth brighter than all the other stars." IGNATIUS, doubtless, had this from those who had actually seen it! PRUDENTIUS (4th cent. A.D.) says that not even the morning star was so fair. Archbishop Trench, who quotes these authorities, says "This star, I conceive, as so many ancients and moderns have done, to have been a new star in the heavens."

One step more places this new star in the constellation of COMA, and with new force makes it indeed "His star"—the "Sign" of His "coming forth from Bethlehem." Will it be "the sign of the Son of Man in heaven" (Matt. xxiv. 30) when He shall "come unto" this world again to complete the wondrous prophecies written of Him in the heavenly and earthly Revelations? *

Thus does the constellation of COMA reveal that the coming "Seed of the woman" was to be a child born, a son given.

But He was to be more: He was to be God and

* It ought also to be noted that in the preceding year there were three conjunctions of the planets Jupiter and Saturn, at the end of May and October, and at the beginning of December. Kepler (1571–1631) was the first to point this out, and his calculations have been confirmed by the highest authorities. These conjunctions occurred in the sign of PISCES: and this sign, according to all the ancient Jewish authorities (Josephus, Abarbanel, Eliezer, and others), has special reference to *Israel.* The conjunction of Jupiter and Saturn, they hold, always marked the occurrence of some event *favourable to Israel;* while Kepler, calculating backwards, found that this astronomical phenomenon always coincided with some great historical crisis, viz.: the Revelation to Adam, the birth of Enoch, the Revelation to Noah, the birth of Moses, the birth of Cyrus, the birth of Christ, the birth of Charlemagne, and the birth of Luther.

man—two natures in one person! This is the lesson of the next picture.

§ 2. CENTAURUS (The Centaur).

The Despised Sin-offering.

It is the figure of a being with two natures. Jamieson, in his *Celestial Atlas*, 1822, says, "On the authority of the most accomplished Orientalist of our own times, the Arabic and Chaldaic name of this constellation is בזה." Now this Hebrew word *Bezeh* (and the Arabic *Al Beze*) means *the despised.* It is the very word used of this Divine sufferer in Isa. liii. 3, "He is DESPISED (נִבְזֶה) and rejected of men."

The constellation contains thirty-five stars. Two of the 1st magnitude, one of the 2nd, six of the 3rd, nine of the 4th, etc., which, together with the four bright stars in the Cross make a brilliant show in southern latitudes.

The brightest star, α (in the horse's fore-foot), has come down to us with the ancient name of *Toliman*, which means *the heretofore and hereafter*, marking Him as the one "which is, and which was, and which is to come—the Almighty" (Rev. i. 8). Sir John Herschell observed this star to be growing rapidly brighter. It may be, therefore, one of the changeable stars, and its name may be taken as an indication of the fact that it was known to the ancients.

Another name for the constellation was in Hebrew, *Asmeath*, which means a *sin-offering* (as in Isa. liii. 10).

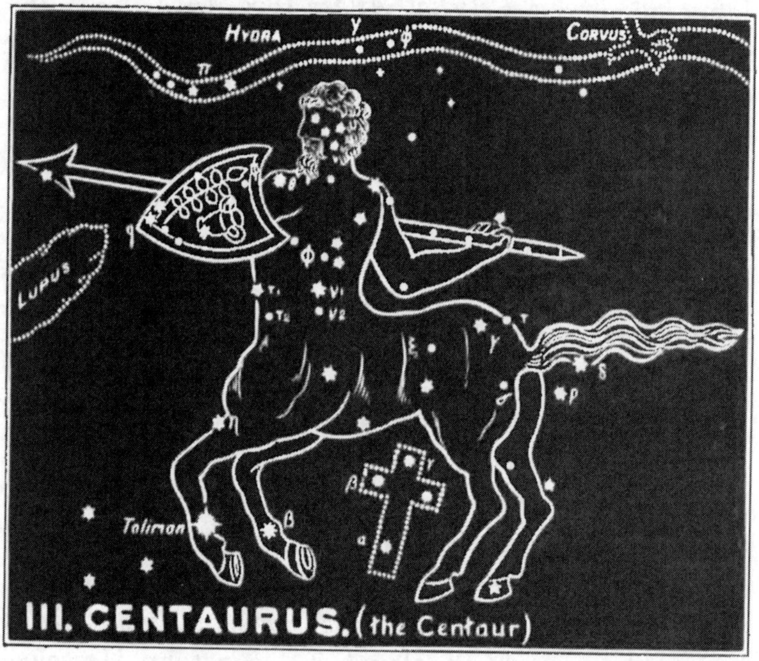

III. CENTAURUS. (the Centaur)

The Greek name was *Cheiron*, which means *the pierced*, or *who pierces*. In the Greek fables *Cheiron* was renowned for his skill in hunting, medicine, music, athletics, and prophecy. All the most distinguished heroes of Greece are described as his pupils. He was supposed to be immortal, but he voluntarily agreed to die; and, wounded by a poisoned arrow (not intended for him) while in conflict with a wild boar, he transferred his immortality to Prometheus; whereupon he was placed amongst the stars.

We can easily see how this fable is the ignorant perversion of the primitive Revelation. The true tradition can be seen dimly through it, and we can discern Him of whom it spoke,—the all-wise, all-powerful Teacher and Prophet, who "went about doing good," yet "despised and rejected of men," laying down His life that others might live.

It is one of the lowest of the constellations, *i.e.* the farthest south from the northern centre. It is situated immediately over the Cross, which bespeaks His own death; He is seen in the act of destroying the enemy.

Thus these star-pictures tell us that it would be as a *child* that the *Promised Seed* should come forth and grow and wax strong in spirit and be filled with wisdom (Luke ii. 40); and that as a man having two natures He should suffer and die. Then the third and last section in this first chapter of this First Book goes on to tell of His second coming in glory.

§ 3. BOÖTES (The Coming One).

He cometh.

This constellation still further develops this wondrous personage.

He is pictured as a man walking rapidly, with a spear in his right hand and a sickle in his left hand.

The Greeks called him *Bo-ö-tes*, which is from the Hebrew root *Bo* (בּוֹא, *to come*), meaning *the coming*. It is referred to in Ps. xcvi. 13:—

> "For He cometh,
> For He cometh to judge the earth;
> He shall judge the world in righteousness,
> And the people with His truth."

It it probable that his ancient name was *Arcturus* * (as referred to in Job ix. 9), for this is the name of the brightest star, α (in the left knee). *Arcturus* means *He cometh.*†

The ancient Egyptians called him *Smat*, which means *one who rules*, *subdues*, and *governs*. They also called him *Bau* (a reminiscence of the more ancient *Bo*), which means also *the coming one.*

---

* The ancient name could not have been *Boötes!* though it is derived from, and may be a reminiscence of the Hebrew.

† ARATUS calls him *Arctophylax*, *i.e.*, the guardian of Arctos, the flock of the greater fold, called to-day the Great Bear:—

> "Behind, and seeming to urge on the Bear,
> Arctophylax, on earth Boötes named,
> Sheds o'er the Arctic car his silver light."

By some moderns he is mistakenly called *The Waggoner*. Hence the allusion of Thompson:—

> "Wide o'er the spacious regions of the North,
> Boötes urges on his tardy wain."

This perversion scarcely does justice even to human common sense, as waggoners do not use a sickle for a whip!

IV. BOÖTES (the Coming One)

The star μ (in the spear-head) is named *Al Katurops*, which means *the branch, treading under foot.*

The star ϵ (just below the waist on his right side) is called *Mirac*, or *Mizar*, or *Izar*. *Mirac* means *the coming forth as an arrow; Mizar*, or *Izar*, means *the preserver, guarding.*

The star η is called *Muphride, i.e. who separates.*

The star β (in the head) is named *Nekkar, i.e. the pierced* (Zech. xii. 10), which tells us that this coming judge is the One who was pierced Another Hebrew name is *Merga, who bruises.**

This brings us back again to Gen. iii. 15, and closes up this first chapter of the First Book (VIRGO). It shows us the *Person* of the Promised Seed from the beginning to the end, from the first promise of the birth of the Child in Bethlehem, to the final coming

---

* The constellation is a very brilliant one, having 54 stars, viz., one of the 1st magnitude, six of the 3rd, eleven of the 4th, etc.

The constellation of the *Canes Venatici* (*the Greyhounds*), *i.e.*, the two dogs (Asterion and Chara), which Boötes holds by a leash, is quite a modern invention, being added by Hevelius (1611–1687). The bright star of the 3rd magnitude in the neck of Chara, was named "*Cor Caroli*" (*the heart of Charles*) by Sir Charles Scarborough, physician to Charles II., in honour of Charles I., in 1649. This is a good example of the almost infinite distance between the ancient and modern names. The former are full of mysterious significance and grandeur, while the latter are puerile in the extreme, almost approaching to the comic! *e.g.*, the Air Pump, the Painter's Easel, the Telescope, the Triangle, the Fly, the Microscope, the Indian, the Fox and Goose, the Balloon, the Toucan (or American Goose), the Compasses, Charles's Oak, the Cat, the Clock, the Unicorn, &c. The vast difference can be at once seen between those designed by the ancients and those added by astronomers in more recent times.

These new constellations were added, 22 by Hevelius (1611–1687); and 15 by Halley (1656–1742). They were formed for the purpose of embracing those stars which were not included in the ancient constellations. This shows that the old constellations were not designed, like the modern ones, merely for the sake of enabling astronomers to identify the positions of particular stars. In this case *all* the stars would have been included. *The object was exactly the opposite!* Instead of the pictures being designed to serve to identify the stars, only certain stars were used for the purpose of helping *to identify the pictures!*

This is another important proof of the truth of our whole argument.

of the great Judge and Harvester to reap the harvest of the earth. This was the vision which was afterwards shown to John (Rev. xiv. 15, 16), when he says, "I looked; and behold a white cloud, and upon the cloud one sat like unto the Son of Man, having on His head a golden crown, and in His hand a sharp sickle. And another angel came out of the temple, crying with a loud voice to Him that sat on the cloud, Thrust in thy sickle and reap; for the time is come for Thee to reap; for the harvest of the earth is ripe. And He that sat on the cloud thrust in His sickle on the earth; and the earth was reaped."

This is the conclusion of the *first chapter* of this First Book. Here we see the woman whose Seed is to bruise the serpent's head, the Virgin-Born, the Branch of Jehovah, perfect man and perfect God, Immanuel, "God with us," yet despised and rejected of men, and yielding up His life that others may have life for evermore. But we see Him coming afterwards in triumphant power to judge the earth.

This is only one chapter of this First Book, but it contains the *outline* of the whole volume, complete in itself, so far as it regards the Person of the Coming One. Like the Book of Genesis, it is the seed-plot which contains the whole, all the rest being merely the development of the many grand details which are included and shut up within it. It is only one chapter out of twelve, but it distinctly foreshadows the end—even "the sufferings of Christ and the glory which should follow."

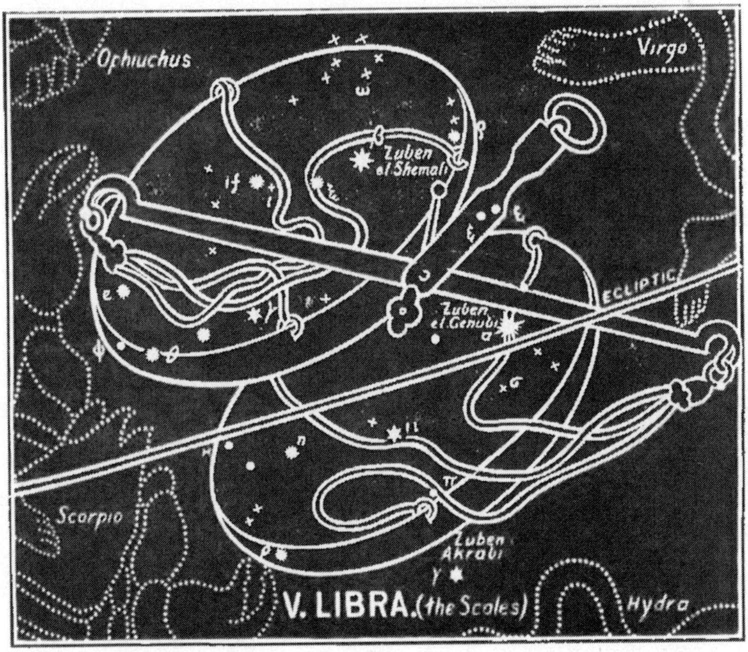

V. LIBRA. (the Scales)

CHAPTER II.

The Sign LIBRA.

The Redeemer's Atoning Work; or *The Price deficient balanced by the Price which covers.*

In the first chapter of this book we saw that this Coming Seed of the woman was, among other things, to give up His life for others.

The *second* chapter is going to define and develope the manner and object of this death.

The name of the Sign, together with its three constellations and the names of the stars composing them, give the complete picture of this Redemption.

The Sign contains 51 stars, two of which are of the 2nd magnitude, one of the 3rd, eight of the 4th, etc.

The Hebrew name is *Mozanaim, the Scales, weighing*. Its name in Arabic is *Al Zubena, purchase*, or *redemption*. In Coptic, it is *Lambadia, station of propitiation* (from *Lam, graciousness*, and *badia, branch*). The name by which it has come down to us is the Latin, *Libra*, which means *weighing*, as used in the Vulgate (Isa. xl. 12).

Libra contains three bright stars whose names supply us with the whole matter. The brightest, α (in the lower scale), is named *Zuben al Genubi*, which means *the purchase*, or *price which is deficient*. This points to the fact that man has been utterly ruined. He is "weighed in the balances and found wanting."

> "None of them can by any means redeem his brother,
> Nor give to God a ransom for him;
> For the redemption of their soul is costly,
> And must be let alone for ever."
> (Ps. xlix. 7, R.V.)

> "Surely men of low degree are vanity (Heb. *a breath*),
> And men of high degree are a lie;
> In the balances they go up;
> They are altogether lighter than vanity" (Heb. *a breath*).
> (Ps. lxii. 9, R.V.)

This is the verdict pronounced and recorded by this star *Zuben al Genubi*.

Is there then no hope? Is there no one who can pay the price?

Yes; there is "the Seed of the woman." He is not merely coming as a child, but He is coming as an atoning sacrifice.

He is coming for the purpose of Redemption! He can pay *the price which covers!* Hence in the upper scale we have another bright star with this very name *Zuben al Chemali*—THE PRICE WHICH COVERS! Praised be God! "They sang a new song, saying, Thou art worthy . . . for Thou wast slain, and hast redeemed us to God by thy blood." (Rev. v. 9.) This is the testimony of β, the second brightest star! It has another name, *al Gubi*, *heaped up*, or *high*, telling of the infinite value of this redemption price. But there is a third star, γ, below, towards *Centaurus* and the *Victim* slain, telling, by that and by its name, of the *conflict* by which that redemption would be accomplished. It is called *Zuben Akrabi* or *Zuben al Akrab*, which means *the price of the conflict!*

There is, however, some reason to suppose that Libra is a very ancient Egyptian corruption, bringing in human merit instead of Divine righteousness; "the way of Cain" instead of the way of God. In the more ancient Akkadian the months were called after the names of the signs,* and the sign of the seventh month is the sign that we now call Libra. The Akkadian name for it was *Tulku*. *Tul* means *mound* (like *dhul* and *dul*), and *ku* means *sacred;* hence, *Tulku* means *the sacred mound*, or *the holy altar*.†

Not only is the name and its meaning different, but the teaching is infinitely greater and more important, if we may believe that the original picture of this sign was not a pair of scales, but the representation of *a holy altar*. This would agree still better with the three constellations which follow.

The names of the stars would also be more appropriate, for it is the Sacrifice of Christ which they foreshadowed, and here it was that the price which covered was paid, and outweighed the price which was deficient. What that price was to be, and how it was to be paid, and what was to be the result in the Person of the Redeemer, is set forth in detail in the three sections of this chapter by the constellations of *The Cross* endured, *The Victim* slain, and *The Crown* bestowed.

---

* See quotation from Dr. Budge, on page 12.

† And certainly the symbol by which it is still known ♎ is more like the top of an altar (See *Ara*, Plate XIV.) than a pair of balances, to which we can trace no resemblance whatever. See Note in the Appendix.

§ 1. CRUX (The Cross).

The Cross Endured.

The Hebrew name was *Adom*, which means *cutting off*, as in Dan. ix. 26:—"After threescore and two weeks shall Messiah be cut off." The last letter of the Hebrew alphabet was called *Tau*, which was anciently made in the form of a cross. The ancient *Phœnician* was 𐤕; the ancient *Hebrew*, as found on coins, was × and +; the *Aramaic*, as found on Egyptian monuments, was a transition p or h, which passed into the present square Hebrew character ת. This letter is called *Tau*, and means *a mark;* especially *a boundary-mark, a limit or finish.* And it is the last letter, which *finishes* the Hebrew alphabet to this day.

The Southern Cross was just visible in the latitude of Jerusalem at the time of the first coming of our Lord to die. Since then, through the gradual recession of the Polar Star, it has not been seen in northern latitudes. It gradually disappeared and became invisible at Jerusalem when the Real Sacrifice was offered there; and tradition, which preserved its memory, assured travellers that if they could go far enough south it would be again seen. Dante sang of "the four stars never beheld but by the early race of men." It was not until the sixteenth century had dawned that missionaries and voyagers, doubling the Cape for the first time, and visiting the tropics and southern seas, brought back the news of "a

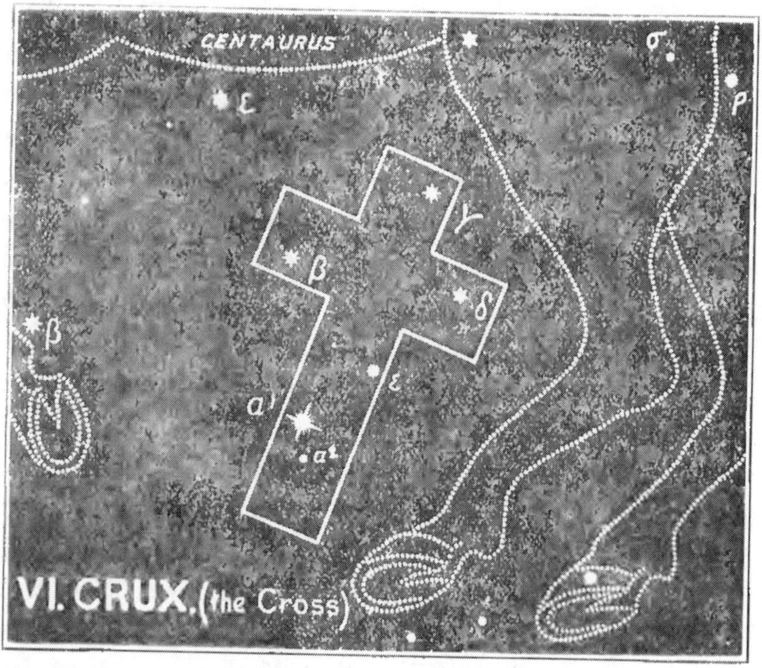
CENTAURUS
VI. CRUX. (the Cross)

wonderful cross more glorious than all the constellations of the heavens."

It is a small asterism, containing only about five stars, *viz.*, one of the 1st magnitude, two of the 2nd, one of the 3rd, and one of the 4th. Four of these are in the form of a cross.

Long before the Christian Era this sign of the Cross had lost its true meaning, and had been perverted in Babylon and Egypt as it has since been desecrated by Rome. The Persians and Egyptians worshipped it. The cakes made and eaten in honour of the Queen of Heaven were marked with it. This heathen custom Rome has adopted and adapted in her Good Friday cakes, which are thus stamped. But all are alike ignorant of what it means, *viz.*, "IT IS FINISHED."

In Egypt, and in the earliest times, it was the sign and symbol of *life*. To-day, Romanists use it as the symbol of *death!* But it means *life!* Natural life given up, and eternal life procured. Atonement, finished, perfect, and complete; never to be repeated, or added to. All who partake of its benefits in Christ now, in grace, by faith "ARE made nigh by the blood of Christ" (Eph. ii. 13), and of them Jesus says, "He that heareth my voice, and believeth on Him that sent me HATH everlasting life, and shall not come into judgment; but IS PASSED from death unto life" (John v. 24). So perfect and complete is the work which Jesus finished on the Cross that we cannot seek to add even our repentance, faith, tears, or prayers, without practically asserting that the work of Christ is not finished, and is not sufficient!

The Hebrew names of this constellation—*Adom* and *Tau*—rebuke our Pharisaic spirit, which is the relic and essence of all false religions, and points to the blessed fact that the Sacrifice was offered "once for all," and the atoning work of Redemption completely finished on Calvary.

> "'Tis finished! the Messiah dies!
> Cut off for sins, but not His own;
> Accomplished is the sacrifice,
> The great redeeming work is done."

In the ancient Egyptian Zodiac of Denderah this first Decan of LIBRA is represented as a lion with his tongue hanging out of his mouth, as if in thirst, and a female figure holding a cup out to him. Under his fore feet is the hieroglyphic symbol of running water. What is all this but "the Lion of the tribe of Judah" brought down "into the dust of death," and saying "I am poured out like water . . . my strength is dried up" (Ps. xxii. 13–18): "I thirst" (John xix. 28): "and in my thirst they gave me vinegar to drink" (Ps. lxix. 21)?

The Egyptian name of this Lion, however, points to his ultimate triumph, for it is called *Sera*, that is, *victory!*

This brings us to—

§ 2. LUPUS OR VICTIMA (The Victim).

The Victim Slain.

Its modern name is *Lupus* (a wolf), because it looks like one. It may be any animal. The great point

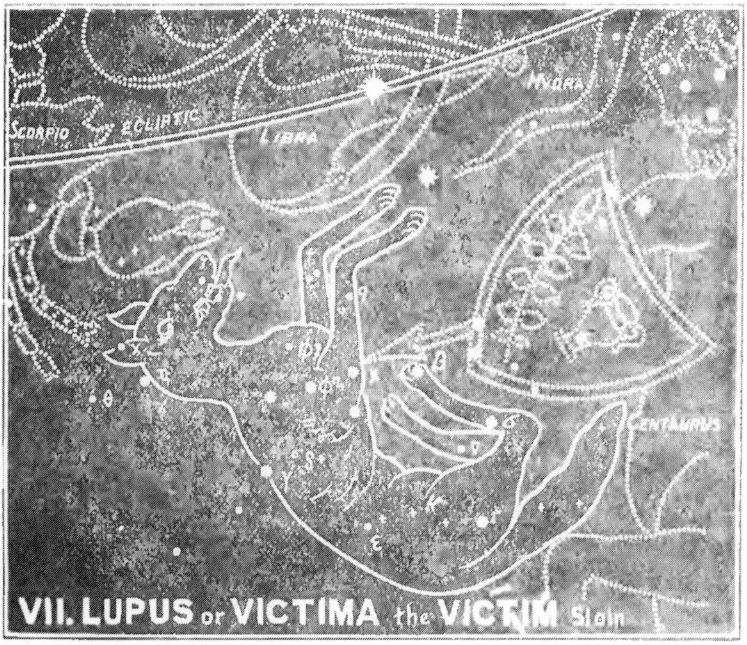

VII. LUPUS or VICTIMA the VICTIM Slain

of this ancient constellation is that the animal has been *slain*, and is in the act of falling down dead.

Its Greek name is *Thera*, *a beast*, and *Lycos*, *a wolf.* Its Latin name is *Victima*, or *Bestia* (Vulg. Gen. viii. 17), which sufficiently indicates the great lesson. This is confirmed by its ancient Hebrew name, *Asedah*, and Arabic *Asedaton*, which both mean *to be slain.*

More than 22 of its stars have been catalogued. None of them are higher than the 4th magnitude; most of them are of the 5th or 6th.

True, He was "by wicked hands crucified and slain," but He is slain here by the Centaur, *i.e.* by Himself! To make it perfectly clear that it was His own act (without which His death would lose all merit), He uttered those solemn words: "I lay down my life for the sheep. . . . No man taketh it from me, but I lay it down of myself. I have power to lay it down, and I have power to take it again" (John x. 15–18). He "offered Himself without spot to God." "He put away sin by the sacrifice of Himself" (Heb. ix. 11, 26).

In the ancient Zodiac of Denderah He is pictured as a little child with its finger on its lips, and He is called *Sura*, *a lamb!* In other pictures He has, besides, the horn of a goat on one side of His head. All this pointed to one and the same great fact, *viz.*, the development and explanation of what was meant by *the bruising of His heel!* It meant that this Promised

D 2

Seed of the woman should come as a child, that He should suffer, and die upon the Cross, for

> "He was brought as a lamb to the slaughter;
> And as a sheep before her shearers is dumb;
> SO HE opened not his mouth."
>
> (Isa. liii. 7.)

Hence, the constellation prefigures a silent, willing sacrifice—Christ Jesus, who, "being found in fashion as a man, humbled Himself, and became obedient unto death, even the death of the Cross" (Phil. ii. 5–8).

§ 3. CORONA (The Crown).

The Crown Bestowed.

"Wherefore God also hath highly exalted Him, and given Him a name which is above every name, that at the name of Jesus every knee should bow."

This is what is foreshown by this concluding section of the second chapter. Each chapter ends with glory. As in the written Word of God, we frequently have the glory of the Second Coming mentioned without any allusions to the sufferings of the First Coming, but we never have the First Coming in humiliation mentioned without an immediate reference to the glory of the Second Coming.

So here, the Cross is closely followed by the Crown! True, "we see not yet all things put under Him, but we see Jesus . . . for the suffering of death crowned with glory and honour (Heb. ii. 9).

Yes, "the crowning day is coming," and all heaven shall soon resound with the triumphant song, "Thou

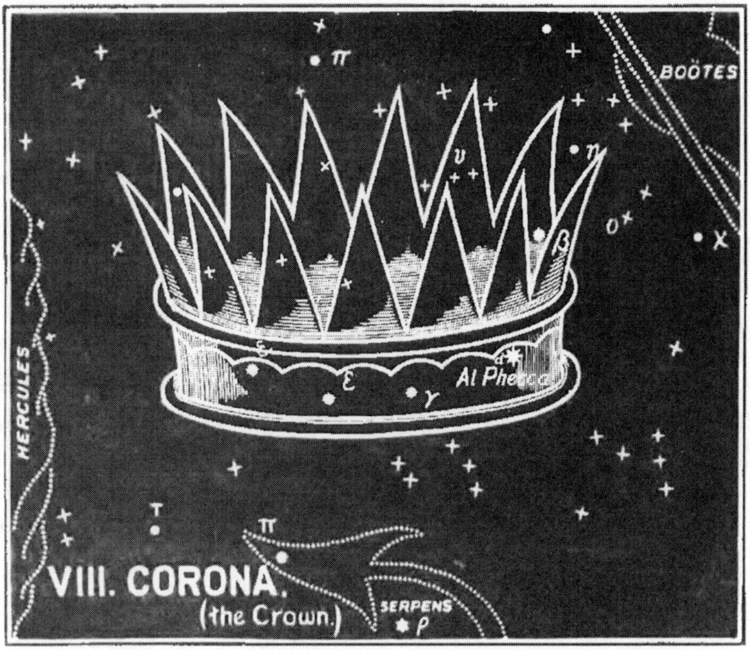
BOÖTES
HERCULES
Al Phecca
SERPENS
VIII. CORONA.
(the Crown.)

art worthy, . . . for Thou wast slain and hast redeemed us to God by Thy blood" (Rev. v. 9).

The shameful Cross will be followed by a glorious crown, and "every tongue shall confess that Jesus Christ is Lord, to the glory of God the Father."

"Mighty Victor, reign for ever,
  Wear the crown so dearly won;
Never shall Thy people, never
  Cease to sing what Thou hast done.
Thou hast fought Thy people's foes;
Thou wilt heal Thy people's woes!"

The Hebrew name for the constellation is *Atarah*, *a royal crown*, and its stars are known to-day in the East by the plural, *Ataroth!*

Its Arabic name is *Al Iclil*, *an ornament*, or *jewel*.

It has 21 stars: one of the 2nd magnitude and six of the 4th. It is easily known by the stars θ, β, α, γ, δ, and ϵ, which form a crescent.

Its brightest star, α, has the Arabic name of *Al Phecca*, *the shining*.

Thus ends this solemn chapter of LIBRA, which describes the great work of Redemption, beginning with the Cross and ending with the Crown. The Redeemer's work of Atonement is most blessedly set forth, and He alone is seen as the substitute for lost sinners.

"What wondrous love, what mysteries
  In this appointment shine!
My breaches of the law are His,
  And His obedience mine."

CHAPTER III.

The Sign SCORPIO.

The Redeemer's Conflict.

We come now right into the heart of the conflict. The star-picture brings before us a gigantic scorpion endeavouring to sting in the heel a mighty man who is struggling with a serpent, but is crushed by the man, who has his foot placed right on the scorpion's heart.

The Hebrew name is *Akrab*, which is the name of a scorpion, but also means *the conflict*, or *war*. It is this that is referred to in Ps. xci. 13:

> "Thou shalt tread upon the lion and adder.
> The young lion and the dragon shalt thou trample
> under feet."

David uses the very word in Ps. cxliv. 1, where he blesses God for teaching his hands *to war*.

The Coptic name is *Isidis*, which means *the attack of the enemy*, or *oppression*; referring to "the wicked that oppress me, my deadly enemies who compass me about" (Ps. xvii. 9).

The Arabic name is *Al Akrab*, which means *wounding him that cometh*.

There are 44 stars altogether in this sign. One is of the 1st magnitude, one of the 2nd, eleven of the 3rd, eight of the 4th, etc.

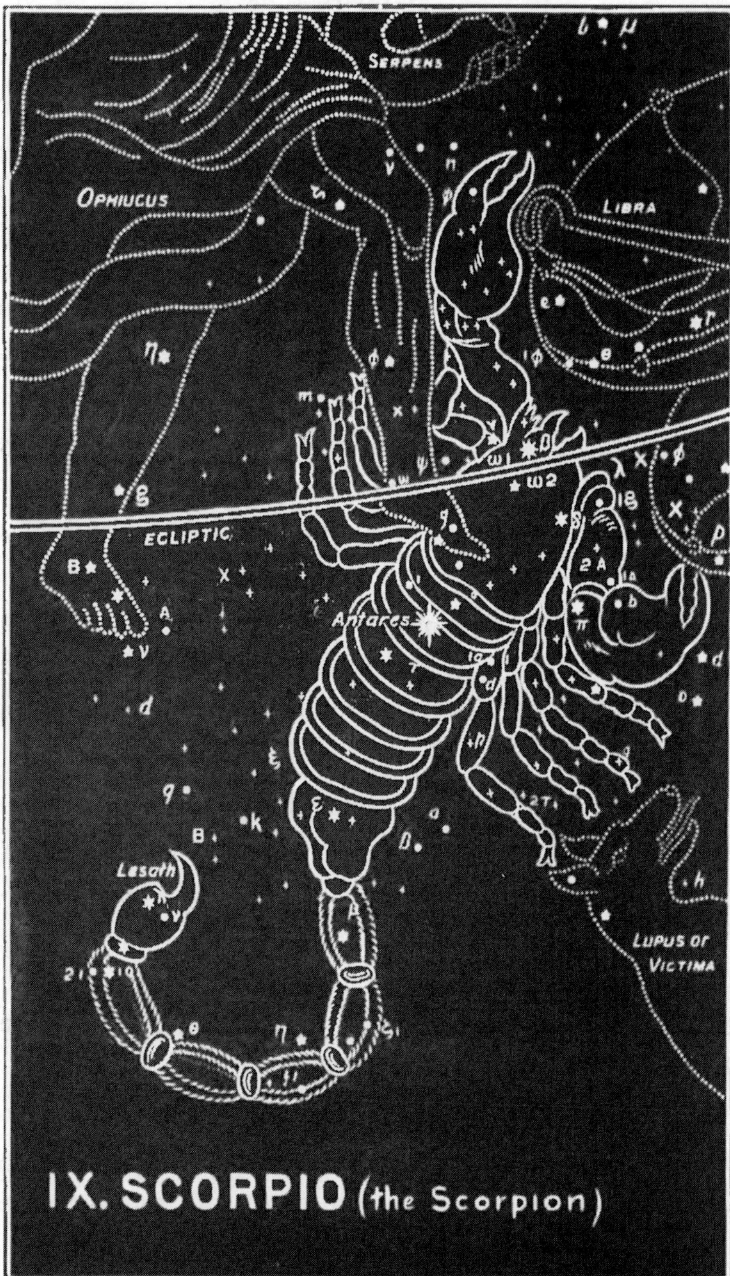
SERPENS
OPHIUCUS
LIBRA
ECLIPTIC
Antares
Lesath
LUPUS or
VICTIMA
IX. SCORPIO (the Scorpion)

The brightest star, α (in the heart), bears the ancient Arabic name of *Antares*, which means *the wounding*. It is called by the Latins *Cor Scorpii*, because it marks the scorpion's heart. It shines ominously with a deep red light. The sting is called in Hebrew *Lesath* (Chaldee, *Lesha*), which means *the perverse*. The stars in the tail are also known as *Leshaa*, or *Leshat*.*

The scorpion is a deadly enemy (as we learn from Rev. ix), with poison in its sting, and all the names associated with the sign combine to set forth the malignant enmity which is "set" between the serpent and the woman's Seed.

That enmity is shown more fully in the written Word, where we see the attempt of the enemy (in Exod. i.) to destroy every male of the seed of Abraham, and how it was defeated.

We see his effort repeated when he used Athaliah to destroy "all the seed royal" (2 Kings xi.), and how "the king's son" was rescued "from among" the slain.

We see his hand again instigating Haman, "the Jews' enemy," to compass the destruction of the whole nation, but defeated in his designs.

When the woman's Seed, the virgin's Son, was born, we are shown the same great enemy inciting Herod to slay all the babes in Bethlehem (Matt. ii.), but again he is defeated.

* *Antares* seems also to have been known as *Lesath*.

In the wilderness of Judæa, and in the Garden of Gethsemane the great conflict is renewed. "This is your hour and the power of darkness,"* He said to His enemies.

The real wounding in the heel was received at the Cross. It was there the scorpion struck the woman's seed. He died, but was raised again from the dead "to destroy the works of the devil."

To show us this; to prevent any mistake; to set forth the fact that this conflict only *apparently* ended in defeat, and that it did not really so end, we have the first two constellations belonging to this sign presented *in one picture!* Indeed, the picture is threefold, for it includes the sign itself (as shown on the cover)!

If these pictures had been separated, then the conflict would have been separated from the victory; the deadly wound of the serpent's head from the temporary wound in the Victor's heel. Hence, *three* pictures are required, in which the *scorpion*, the *serpent*, and the *man*, are all involved, in order to present at the same time the triumphant issue of the conflict.

Hence, we must present, and consider together, the first two sections of this mysterious chapter.

§§ 1 AND 2. SERPENS AND OPHIUCHUS.

The Struggle with the Enemy.

Here, *Serpens*, the serpent, is seen struggling vainly in the powerful grasp of the man who is named *O-phi-u-chus*. In Latin he is called Serpentarius.

* Luke xxii. 53: comp. Col. i. 13 and Eph. vi. 12.

X
SERPENS (the Serpent)
OPHIUCHUS (the Serpent Holder)
Cerberus &
Ramus Pomifer
HERCULES
Boötes
VIRGO
LIBRA
Zuben Akrabi
Zuben Al Chamali
Yed
Unuk
SCORPIO
Saiph
ECLIPTIC
SAGITTARIUS

He is at one and the same moment shown to be seizing the serpent with his two hands, and treading on the very heart of the scorpion, marked by the deep red star *Antares* (wounding).

Just as we read the first constellation of the woman and child *Coma*, as expounding the first sign VIRGO, so we have to read this first constellation as expounding the second sign LIBRA. Hence, we have here a further picture, showing the object of this conflict on the part of the scorpion.

In Scorpio we see merely the effort to wound *Ophiuchus* in the heel; but here we see the effort of the serpent to seize THE CROWN, which is situated immediately over the serpent's head, and to which he is looking up and reaching forth.

The contest is for Dominion! It was the Devil, in the form of a serpent, that robbed the first man of his crown; but in vain he struggled to wrest it from the sure possession of the Second Man. Not only does he fail in the attempt, but is himself utterly defeated and trodden under foot.

There are no less than 134 stars in these two constellations. Two are of the 2nd magnitude, fourteen of the 3rd, thirteen of the 4th, etc.

The brightest star in the Serpent, α (in the neck), is named *Unuk*, which means *encompassing*. Another Hebrew name is *Alyah, the accursed*. From this is *Al Hay* (Arabic), *the reptile*. The next brightest star is β (in the jaw), named, in Arabic, *Cheleb*, or *Chelbalrai, the serpent enfolding*. The Greek name,

Ophiuchus, is itself from the Hebrew and Arabic name *Afeichus*, which means *the serpent held*. The brightest star in *Ophiuchus*, α (in the head), is called *Ras al Hagus* (Arabic), *the head of him who holds*.

Other Hebrew names of stars, not identified, are *Triophas*, *treading under foot; Saiph* (in the foot* of Ophiuchus), *bruised; Carnebus, the wounding; Megeros, contending*.† In the Zodiac of Denderah we have a throned human figure, called *Api-bau, the chief who cometh*. He has a hawk's head to show that he is the enemy of the serpent, which is called *Khu*, and means *ruled* or *enemy*.

All these combine to set before us in detail the nature of the conflict and its final issue. That final issue is, however, exhibited by the last of the three constellations of this chapter. The Victor Himself requires a whole picture to fully set forth the glorious victory. This brings us to—

§ 3. HERCULES (The Mighty Man).

The Mighty Vanquisher.

Here the mighty one, who occupies a large portion of the heavens, is seen bending on one knee, with his right heel lifted up as if it had been wounded,

---

* In 1604 a new star appeared in the eastern foot of Ophiuchus, but disappeared again in 1605.

† There is an ancient Greek fable which calls Ophiuchus Æsculapius, the son of Apollo. Having restored Hippolytus to life, he was everywhere worshipped as the god of health, and hence the serpent entwined around him is, to this day, the symbol of the medical art! This, however, is, doubtless, another perversion of the primitive truth that the Coming One in overcoming the serpent, should become the great healer of all the sorrows of the world, and cause all its groanings to cease.

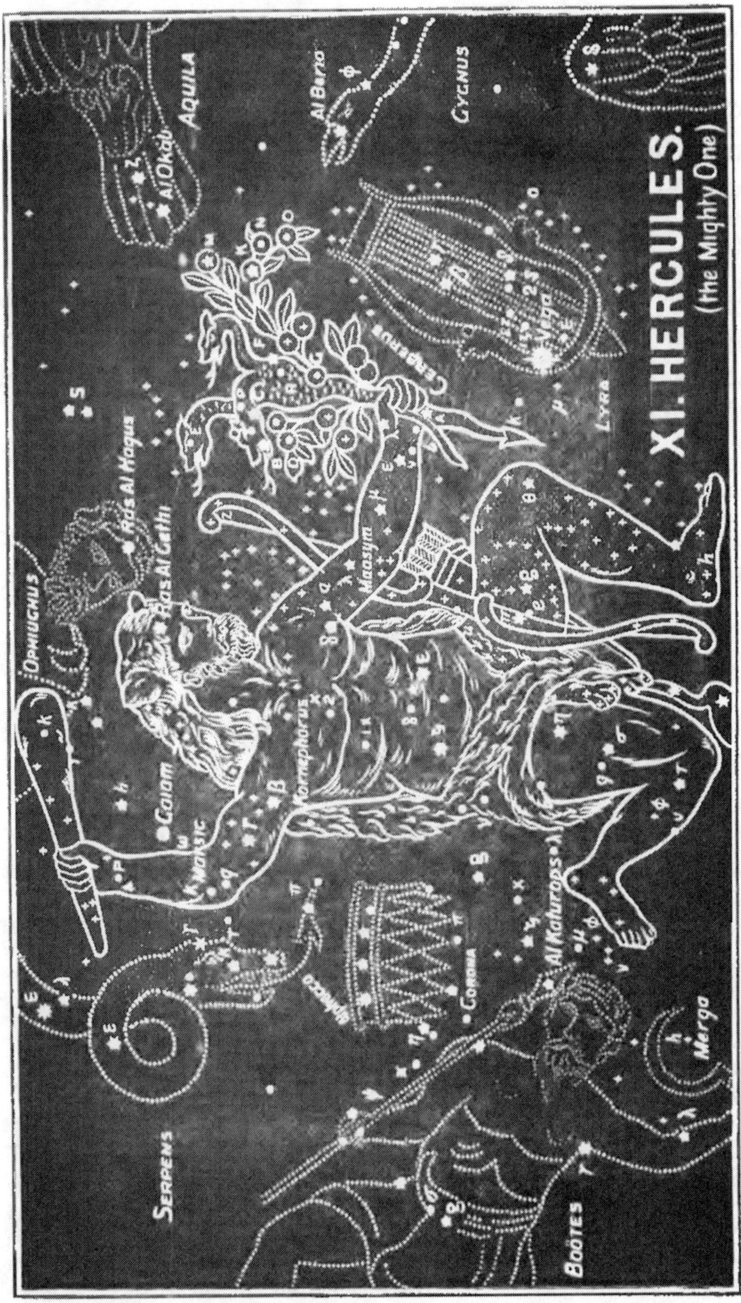
XI. HERCULES.
(the Mighty One)
AQUILA
Al Okab
Al Bario
CYGNUS
LYRA
Vega
CERBERUS
Ras Al Hagus
Ras Al Gethi
OPHIUCHUS
Maasym
Kornephorus
Caiam
Marsic
Al Katurops
CORONA
SERPENS
BOOTES
Merga

while his left foot is set directly over the head of the great dragon. In his right hand he wields a great club, and in his left hand he grasps a triple-headed monster (*Cerberus*). And he has the skin of a lion, which he has slain, thrown around him.*

In the Zodiac of Denderah we have a human figure, likewise with a club. His name is *Bau*, which means *who cometh*, and is evidently intended for Him who cometh to crush the serpent's head, and "destroy the works of the devil."

In Arabic he is called *Al Giscale*, *the strong one*.

There are 113 stars in this constellation. Seven are of the 3rd magnitude, seventeen of the 4th, etc.

The brightest star, α (in his head), is named *Ras al Gethi*, and means *the head of him who bruises*.

The next, β (in the right arm-pit), is named *Kornephorus*, and means *the branch, kneeling*.

The star κ (in the right elbow) is called *Marsic*, *the wounding*.

The star λ (in the upper part of the left arm) is named *Ma'asyn*, *the sin-offering*.

While ω (in the lower part of the right arm) is *Caiam*, or *Guiam*, *punishing*; and in Arabic, *treading under foot*.

---

* *Cerberus*, or the serpent with three heads, was placed by Hevelius (1611–1687) by the side of Hercules. Bayer had previously placed the apple branch in his hand. This was symbolical of the golden apples of *Hesperides*, which he obtained by killing this three-headed *hydra*, by whom they were guarded. In our picture these are combined, and a bow and quiver added from other ancient authorities.

Thus does everything in the picture combine to set forth the mighty works of this stronger than the strong man armed!

We can easily see how the perversion of the truth by the Greeks came about, and how, when the true foreshadowings of this Mighty One had been lost, the many fables were invented to supply their place. The wiser sort of Greeks knew this perfectly well. ARISTOTLE (in his *Metaphysics*, x. 8) admits, with regard to Greek mythology, that religion and philosophy had been lost, and that much had been "added after the mythical style," while much had come down, and "may have been preserved to our times as the remains of ancient wisdom." Religion, such as it was (POLYBIUS confesses), was recognised as a "necessary means to political ends." NEANDER says that it was "the fragments of a tradition, which transmitted the knowledge of divine things possessed in the earliest times."

ARATUS shews the same uncertainty as to the meaning of this Constellation of *Hercules*. He says:

"Near this, and like a toiling man, revolves
A form. Of it can no one clearly speak,
Nor what he labours at. They call him simply
'The man upon his knees': In desperate struggle
Like one who sinks, he seems. From both his shoulders
His arms are high-uplifted and out-stretched
As far as he can reach; and his right foot
Is planted on the coiléd Dragon's head."

Ancient authorities differ as to the personality of Hercules, and they disagree as to the number, nature, and order of what are sometimes called "the twelve labours of Hercules." But there is no doubt as to the mighty foretold works which the woman's Seed should perform.

From first to last Hercules is seen engaged in destroying some malignant foe: now it is the Nemean lion; then it is the slaying of the boar of Erymanthus; again, it is the conquest of the bull of Crete; then the killing of the three-headed hydra, by whose venom Hercules afterwards died. In the belly of the sea monster he is said to have remained "three days and three nights." This was, doubtless, a perversion of the type of Jonah, introduced by LYCOPHRON, who (living at the court of PTOLEMY PHILADELPHUS, under whose auspices the Hebrew Scriptures were translated into Greek) would have known of that Divine miracle, and of its application to the Coming One. Bishop Horsley believed that the fables of the Greek mythology could be traced back to the prophecies of the Messiah, of which they were a perversion from ignorance or design. This is specially true of Hercules. In his apparently impossible tasks of overthrowing gigantic enemies and delivering captives, we can see through the shadow, and discern the pure light of the truth. We can understand how the original star-picture must have been a prophetic representation of Him who shall destroy the Old Serpent and open the way again, not to fabled "apples of gold," but to the "tree of life" itself. He it is who though suffering in the mighty conflict, and brought to His knee, going down even to "the dust of death," shall yet, in resurrection and advent glory, wield His victorious club, subdue all His enemies, and plant His foot on the Dragon's head. For of Him it is written :—

> "Thou shalt tread upon the lion and adder;
> The young lion and the dragon shalt Thou trample under foot." (Ps. xci. 13.)

> "Come, Lord, and burst the captives' chains,
> And set the prisoners free;
> Come, cleanse this earth from all its stains,
> And make it meet for Thee!
>
> Oh, come and end Creation's groans—
> Its sighs, its tears, its blood,
> And make this blighted world again
> The dwelling-place of God."

CHAPTER IV.

THE SIGN SAGITTARIUS.

The Redeemer's Triumph.

THIS is the concluding chapter of the first great book of this Heavenly Revelation; and it is occupied wholly with the triumph of the Coming One, who is represented as going forth "conquering and to conquer."

The subject is beautifully set forth in the written Word (Ps. xlv. 3–5) :—

> "Gird Thy sword upon Thy thigh, O most mighty,
> [*Gird Thyself*] with Thy glory and Thy majesty,
> And in Thy majesty ride prosperously,
> Because of truth, and meekness, and righteousness;
> And Thy right hand shall teach Thee terrible things.
> Thine arrows are sharp in the heart of the King's enemies;
> Whereby the people fall under Thee."

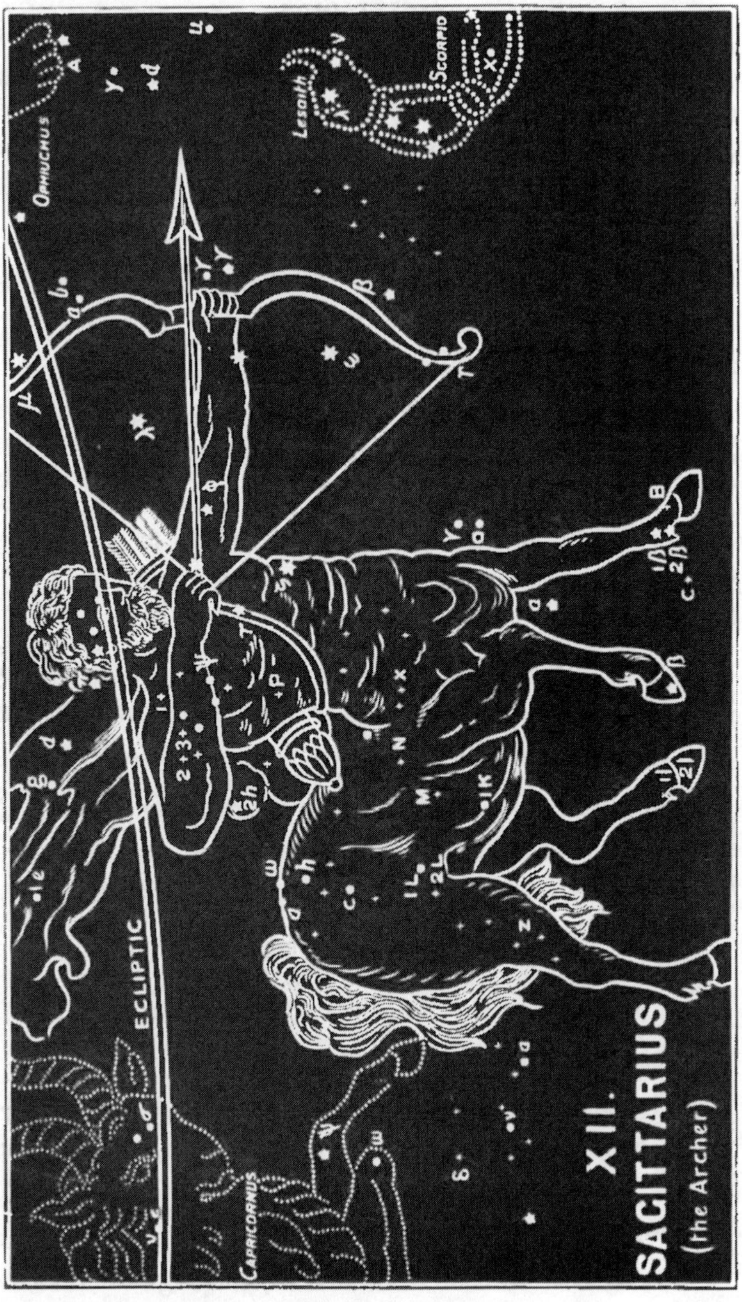

XII.
SAGITTARIUS
(the Archer)

John, in his apocalyptic vision, sees the same mighty Conqueror going forth. "I saw (he says) a white horse, and He that sat on him had a bow, . . . and He went forth conquering and to conquer" (Rev. vi. 2).

This is precisely what is foreshadowed in the star-pictured sign now called by the modern Latin name *Sagittarius*, which means *the Archer.*

The Hebrew and Syriac name of the sign is *Kesith*, which means *the Archer* (as in Gen. xxi. 20). The Arabic name is *Al Kaus*, *the arrow.* In Coptic it is *Pimaere*, *the graciousness*, or *beauty of the coming forth.* In Greek it is *Toxotes*, *the archer*, and in Latin *Sagittarius.*

There are 69 stars in the sign, viz., five of the 3rd magnitude (all in the bow), nine of the 4th, etc.

The names of the brightest stars are significant:—

Hebrew, *Naim*, which means *the gracious one.* This is exactly what is said of this Victor in the same Psalm (xlv.), in the words immediately preceding the quotation above (verse 2):

> "GRACE is poured into Thy lips;
> Therefore God hath blessed Thee for ever."

Hebrew, *Nehushta*, *the going* or *sending forth.*

We see the same in the Arabic names which have come down to us: *Al Naim*, *the gracious one; Al Shaula*, *the dart; Al Warida*, *who comes forth; Ruchba er rami*, *the riding of the bowman.*

An ancient Akkadian name in the sign is *Nun-ki*, which means *Prince of the Earth.*

Again we have the picture of *a Centaur* as to his outward form, *i.e.* a being with two natures. Not now far down in the south, or connected with His sufferings and sacrifice as man; but high up, as a sign of the Zodiac itself, on the ecliptic, *i.e.* in the very path in which the sun "rejoiceth in his going forth as a strong man."

According to Grecian fable, this Sagittarius is *Cheiron*, the chief Centaur; noble in character, righteous in his dealings, divine in his power.

Such will be the coming Seed of the woman in His power and glory:—

> "The sceptre of Thy kingdom is a right sceptre.
> Thou lovest righteousness, and hatest wickedness;
> Therefore God, Thy God, hath anointed Thee with the
> oil of gladness above thy fellows."
>
> (Ps. xlv. 6, 7.)

In the ancient Zodiac of Denderah he is called (as in Coptic) *Pi-macre*, *i.e. graciousness*, *beauty of the appearing* or *coming forth*. The characters under the hind foot read *Knem*, which means *He conquers*.

This is He who shall come forth like as an arrow from the bow, "full of grace," but "conquering and to conquer."

In all the pictures he is similarly represented, and the arrow in his bow is aimed directly at the heart of the Scorpion.

Thus ARATUS sang of *Cheiron*:—

> "'Midst golden stars he stands refulgent now,
> And thrusts the scorpion with his bended bow."

In this Archer we see a faint reflection of Him who shall presently come forth, all gracious, all wise, all powerful; whose arrows shall be "sharp in the heart of the King's enemies."

> "God shall shoot at them with an arrow;
> Suddenly shall they be wounded.
> So they shall make their own tongue to fall upon themselves;
> All that see them shall flee away.
> And all men shall fear, and shall declare the work of God;
> For they shall wisely consider of His doing.
> The righteous shall be glad in the LORD, and shall trust in Him;
> And all the upright in heart shall glory."
>
> (Ps. lxiv. 7–10.)

> "Christ is coming! let Creation
> From her groans and travail cease;
> Let the glorious proclamation
> Hope restore, and faith increase.
> Christ is coming,
> Come, thou blessed Prince of peace."

This brings us to the first of the three constellations or sections of this chapter, which takes up this subject of praise to the Conqueror.

§ 1. LYRA (The Harp).

Praise prepared for the Conqueror.

"Praise waiteth for thee, O God, in Zion" (Ps. lxv. 1). And when the waiting time is over, and the Redeemer comes forth, then the praise shall be given. "We give Thee thanks, O Lord God, the Almighty, which art, and which wast, because thou

S. 1005. E

hast taken to Thee Thy great power, and didst reign" (Rev. xi. 17, R.V.). "Let us be glad and rejoice and give honour unto Him" (Rev. xix. 7). The Twenty-first Psalm should be read here, as it tells of the bursting forth of praise on the going forth of this all-gracious Conqueror.

> "The King shall rejoice in Thy strength, O LORD;
> And in Thy salvation how greatly shall He rejoice! . . .
> Thine hand shall find out all Thine enemies;
> Thy right hand shall find out all that hate thee. . . .
> Their fruit shalt Thou destroy from the earth;
> And their seed from among the children of men.
> For they intended evil against Thee;
> They imagined a mischievous device which they are not able to perform,
> Therefore shalt thou make them turn their back (Heb. *Margin, "set them as a butt"*),
> When Thou shalt make ready Thine arrows upon Thy strings
> [*And shoot them*] against the face of them.
> Be thou exalted, LORD, in thine own strength;
> SO WILL WE SING AND PRAISE THY POWER."
> (Ps. xxi. 1, 8, 10-13.)

Beautifully, then, does *the harp* come in here, following upon the going forth of this victorious Horseman. This Song of the Lamb follows as naturally as does the Song of Moses in Ex. xv. 1: "I will sing unto the LORD, for He hath triumphed gloriously."

Its brightest star, α, is one of the most glorious in the heavens, and by it this constellation may be easily known. It shines with a splendid white lustre. It is called *Vega*, which means *He shall be exalted*. Its root occurs in the opening of the Song

XIII. LYRA. (the Harp)

of Moses, quoted above. Is not this wonderfully expressive?

Its other stars, β and γ, are also conspicuous stars, of the 2nd and 4th magnitude. β is called *Shelyuk*, which means *an eagle* (as does the Arabic, *Al Nesr*); γ is called *Sulaphat*, *springing up*, or *ascending*, as praise.

In the Zodiac of Denderah, this constellation is figured as a hawk or an eagle (the enemy of the serpent) in triumph. Its name is *Fent-kar*, which means *the serpent ruled.*

There may be some confusion between the Hebrew נֶשֶׁר, *Nesher, an eagle*, and עָשׂוֹר, *Gnasor, a harp;** but there can be no doubt about the grand central truth, that praise shall ascend up "as an eagle toward heaven," when "every creature which is in heaven, and on the earth, and such as are in the sea, and all that is in them," shall send up their universal song of praise: "Blessing, and honour, and glory, and power, be unto Him that sitteth upon the throne and unto the Lamb for ever and ever. Amen" (Rev. v. 13, 14).

And for what is all this wondrous anthem of praise? Listen once again. "Alleluia †: Salvation, and glory, and honour, and power, unto the Lord our

* In our picture we have combined the two great thoughts, taking the *harp* from a picture dug up at Herculaneum, and adding an eagle soaring up with it.

† This is the first time that the word "Alleluia" occurs in the New Testament, and it is praise for judgment executed.

Where is its first occurrence in the Old Testament? In Ps. civ. 35, where we have the very same solemn and significant connection:—

"Let the sinners be consumed out of the earth,
And let the wicked be no more.
Bless thou the LORD, O my soul,
HALLELUJAH (Praise ye the LORD)."

God; for TRUE AND RIGHTEOUS ARE HIS JUDGMENTS. . . . And again they said Alleluia" (Rev. xix. 1–3).

> With "that blessed hope" before us,
> Let no HARP remain unstrung;
> Let the coming advent chorus
> Onward roll from tongue to tongue,
> Hallelujah,
> "Come, Lord Jesus," quickly come.

This brings us to—

§ 2. ARA (The Altar).

Consuming Fire Prepared for His Enemies.

Here we have an altar or burning pyre, placed significantly and ominously upside down! with its fires burning and pointing downwards towards the lower regions, called *Tartarus*, or *the abyss*, or "outer-darkness."

It is an asterism with nine stars, of which three are of the 3rd magnitude, four of the 4th, etc. It is south of the Scorpion's tail, and when these constellations were first formed it was visible only on the very lowest horizon of the south, pointing to the completion of all judgment in the lake of fire.

In the Zodiac of Denderah we have a different picture, giving us another aspect of the same judgment. It is a man enthroned, with a flail in his hand. His name is *Bau*, the same name as *Hercules* has, and means *He cometh.* It is from the Hebrew בּוֹא (*Bōh*), *to come*, as in Isa. lxiii. 1:

> "Who is this that cometh from Edom,
> With dyed garments from Bozrah."

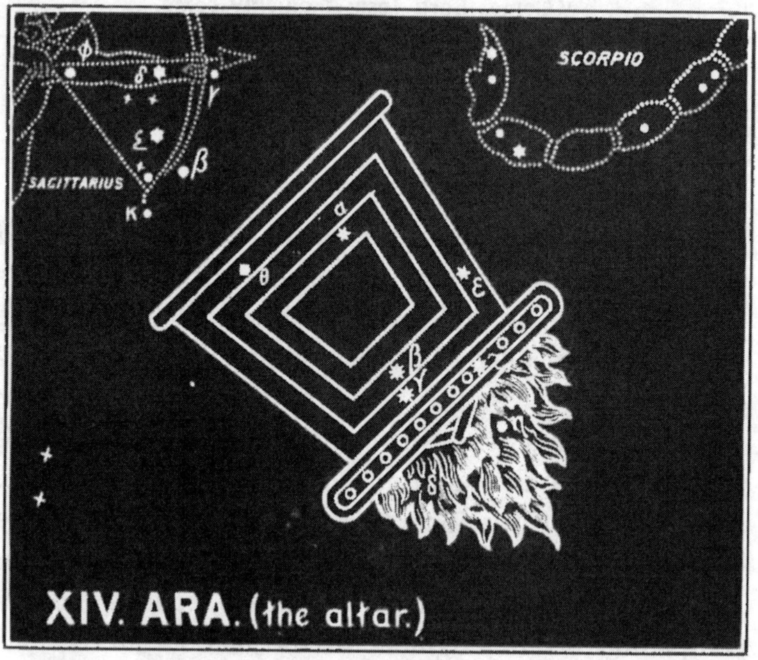

XIV. ARA. (the altar.)

This is a coming in judgment, as is clear from the reason given in verse 4:

> "For the day of vengeance is in Mine heart,
> And the year of My redeemed is come.
> And I looked, and there was none to help;
> And I wondered that there was none to uphold;
> Therefore Mine own arm brought salvation,
> And My fury, it upheld Me."
>
> (Isa. lxiii. 4, 5.)

The completion of judgment, therefore, is what is pictured both by the burning pyre and the Coming One enthroned, with his threshing instrument.

In Arabic it is called *Al Mugamra*, which means *the completing*, or *finishing*. The Greeks used the word *Ara* sometimes in the sense of *praying*, but more frequently in the sense of *imprecation* or *cursing*.

This is the curse pronounced against the great enemy. This is the burning fire, pointing to the *completion* of that curse, when he shall be cast into that everlasting fire "prepared for the devil and his angels." This is the allusion to it written in the midst of the very Scripture from which we have already quoted (p. 66), Ps. xxi., where we read in verse 9 (which we then omitted):—

> "Thou shalt make them as a fiery oven in the time
> of Thine anger:
> The LORD shall swallow them up in His wrath;
> And the fire shall devour them."

This brings us to the final scene, closing up this first great book of the Heavens.

§ 3. DRACO (The Dragon).

The Old Serpent, or the Devil, cast down from Heaven.

Each of the three great books concludes with this same foreshowing of Apocalyptic truth. The same great enemy is referred to in all these pictures. He is the Serpent; he is the Dragon; "the great dragon, that old serpent, called the Devil and Satan" (Rev. xii. 9). The Serpent represents him as the *Deceiver;* the Dragon, as the *Destroyer.*

This *First* Book concludes with the Dragon being cast down from heaven.

The *Second* Book concludes with *Cetus,* the Sea Monster, Leviathan, bound.

The *Third* Book concludes with Hydra, the Old Serpent, destroyed.

Here, at the close of the *First* Book, we see not merely a dragon, but the Dragon *cast down!* That is the point of this great star-picture.

No one has ever seen a dragon; but among all nations (especially in China and Japan), and in all ages, we find it described and depicted in legend and in art. Both Old and New Testaments refer to it, and all unite in connecting with it one and the same great enemy of God and man.

It is against him that the God-Man—"the Son of God—goes forth to war." It is for him that the eternal fires are prepared. It is he who shall shortly be cast down from the heavens preparatory to his completed judgment. It is of him we read, "The great dragon was cast out, that old serpent, called the Devil, and Satan, which deceiveth the whole world: he was cast out and his angels with him. And I heard a loud voice saying in heaven, Now is come salvation, and strength, and the kingdom of our God, and the power of His Christ; for the accuser of our brethren is cast down" (Rev. xii. 9, 10).

It is of him that David sings:—

> "God is my king of old,
> Working salvation in the midst of the earth . . .
> Thou brakest the heads of the dragons in the waters.
> Thou brakest the heads of leviathan in pieces."
> (Ps. lxxiv. 12–14.)

Of him also the Spirit causes Isaiah to say, "In that day, shall this song be sung in the land of Judah";—

> "In that day the LORD, with his sore, and great, and strong sword,
> Shall punish leviathan the piercing (R.V. swift) serpent,
> Even leviathan that crooked serpent;
> And he shall slay the dragon that is in the sea."
> (Isa. xxvi. 1; xxvii. 1.)

This is exactly what is foreshadowed by this constellation of *Draco*. Its name is from the Greek, and means *trodden on*, as in the Septuagint of Ps. xci. 13: "The dragon shalt thou trample under feet," from the Hebrew דָּרַךְ, *Dahrach, to tread.*

In the Zodiac of Denderah it is shown as a serpent under the fore-feet of Sagittarius, and is named *Her-fent*, which means *the serpent accursed!*

There are 80 stars in the constellation; four of the 2nd magnitude, seven of the 3rd magnitude, ten of the 4th, etc.

The brightest star, α (in one of the latter coils), is named *Thuban* (Heb.), *the subtle*. Some 4,620 years ago it was the Polar Star. It is manifest, therefore, that the Greeks could not have invented this constellation, as is confessed by all modern astronomers. It is still a very important star in nautical reckonings, guiding the commerce of the seas, and thus "the god of this world" is represented as winding in his contortions round the pole of the world, as if to indicate his subtle influence in all worldly affairs.

The next star, β (in the head), is called by the Hebrew name *Rastaban*, and means *the head of the subtle (serpent)*. In the Arabic it is still called *Al Waid*, which means *who is to be destroyed*.

The next star, γ (also in the head), is called *Ethanin*, *i.e.*, *the long serpent*, or *dragon*.

The Hebrew names of other stars, not identified, are *Grumian*, *the subtle; Giansar*, *the punished enemy*. Other (Arabic) names are *Al Dib*, *the reptile; El Athik*, *the fraudful; El Asieh*, *the bowed down*.

And thus the combined testimony of every star (without a single exception) of each constellation,

XV. DRAGO. (the Dragon Cast down)

and the constellations of each sign, accords with the testimony of the Word of God concerning the coming Seed of the woman, the bruising of His heel, the crushing of the serpent's head, "the sufferings of Christ, and the glory which should follow."

"From far I see the glorious day,
When He who bore our sins away,
Will all His majesty display.

A Man of Sorrows once He was,
No friend was found to plead His cause,
As all preferred the world's applause.

He groaned beneath sin's awful load,
For in the sinner's place He stood,
And died to bring him back to God.

But now He waits, with glory crowned,
While angel hosts His throne surround,
And still His lofty praises sound.

To few on earth His name is dear,
And they who in His cause appear,
The world's reproach and scorn must bear.

Jesus, Thy name is all my boast,
And though by waves of trouble tossed,
Thou wilt not let my soul be lost.

Come then, come quickly from above,
My soul impatient longs to prove,
The depths of everlasting love."

The Second Book.

THE REDEEMED.

The Result of the Redeemer's Sufferings.

IN the *First* Book we have had before us the work of the Redeemer set forth as it concerned His own glorious person. In this *Second* Book it is presented to us as it affects others. Here we see the *results* of His humiliation, and conflict, and victory—"The sufferings of Christ" and the blessings they procured for His redeemed people.

In Chapter I. we have the Blessings procured.

In Chapter II. their Blessings ensured.

In Chapter III. their Blessings in abeyance.

In Chapter IV. their Blessings enjoyed.

CHAPTER I.

THE SIGN CAPRICORNUS (The Sea Goat).

The Goat of Atonement Slain for the Redeemed.

IT is most noteworthy that this Second Book opens with the Goat, and closes with the Ram: two animals of sacrifice; while the two middle chapters are both connected with fishes.* The reason for this we shall see as we proceed.

Both are combined in the first chapter, or "Sign" of Capricornus.

In all the ancient Zodiacs, or Planispheres, we find a goat with a fish's tail. In the Zodiacs of Denderah and Esneh, in Egypt, it is half-goat and half-fish, and it is there called *Hu-penius*, which means *the place of the sacrifice.*

In the Indian Zodiac it is a goat *passant* traversed by a fish.

There can be no doubt as to the significance of this sign.

In the Goat we have the Atoning Sacrifice, in the Fish we have the people for whom the atonement is made. When we come to the sign "PISCES" we shall see more clearly that it points to the *multitudes* of the redeemed host.

The Goat is bowing its head as though falling down in death. The right leg is folded underneath the body, and he seems unable to rise with the left. The tail of the fish, on the other hand, seems to be full of vigour and life.

The Hebrew name of the sign is *Gedi*, *the kid* or *cut off*, the same as the Arabic *Al Gedi*. CAPRICORNUS is merely the modern (Latin) name of the sign, and means *goat*.

---

* There is a fish tail here. The third Decan of CAPRICORNUS is a fish (*Delphinus*). There is again a fish (*Piscis Australis*) in the next sign (AQUARIUS), and then the following sign is PISCES, or the Fishes. So that the Redeemed Multitudes are presented throughout this Second Book.

There are 51 stars in the sign, three of which are of the 3rd magnitude, three of the 4th, etc. Five are remarkable stars, α and β in the horn and head, and the remaining three, γ, δ, and ε, in the fishy tail. The star α is named *Al Gedi, the kid* or *goat*, while the star δ is called *Deneb Al Gedi, the sacrifice cometh.*

Other star-names in the sign, not identified, are *Dabih* (Syriac), *the sacrifice slain; Al Dabik* and *Al Dehabeh* (Arabic) have the same meaning; *Ma'asad, the slaying; Sa'ad al Naschira, the record of the cutting off.*

Is not this exactly in accord with the Scriptures of truth? There were two goats! Of "the *goat* of the sin-offering" it is written, "God hath given it to you to bear the iniquity of the congregation, to make atonement for them before the LORD" (Lev. x. 16, 17): of the other goat, which was not slain, "he shall let it go into the wilderness" (Lev. xvi. 22). Here is death and resurrection. Christ was "wounded for our transgressions, and bruised for our iniquities." "For the transgression of MY PEOPLE was He stricken" (Isa. liii.). He laid down His life for the sheep.

In the first chapter of the *First* Book we had the same Blessed One presented as "a corn of wheat." Here we see Him come to "die," and hence not abiding alone, but bringing forth "much fruit" (John xii. 24). The living fish proceeds from the dying goat, and yet they form only one body.

XVI. CAPRICORNUS.(the Goat)

That picture, which has no parallel in nature, has a perfectly true counterpart in grace; and "a great multitude, which no man can number," have been redeemed and shall obtain eternal life through the death of their Redeemer.

It is, however, not merely the actual death which is set before us here. The *first* chapter *in each book* has for its great subject *the Person* of the Redeemer in *prophecy* and promise. The *last* chapter in each book has for its subject the fulfilment of that prophecy in victory and triumph, in the Person of the Redeemer: while the two *central* chapters *in each* book are occupied with *the work* which is the accomplishment of the promise, presented in two aspects—the former connected with *grace*, the latter with *conflict*.

Thus the *structure* of each of the three books is an *epanodos*, having for its first and last members the Person of the Redeemer (in "A" in *Prophecy;* in "*A*" in *Fulfilment*), while in the two central members we have the work and its accomplishment (in "B" *in grace;* and in "*B*" *in conflict*).

It may be thus presented to the eye:—

The First Book.

A | VIRGO. The Prophecy of the Bruised Seed.
  B | LIBRA. The work accomplished (in *grace*).
  B | SCORPIO. The work accomplished (in *conflict*).
A | SAGITTARIUS. The fulfilment of the promised victory.

The Second Book.

C | CAPRICORNUS. The Prophecy of the Promised Deliverance.

  D | AQUARIUS. Results of the work bestowed (in grace).

  D | PISCES. Results of the work enjoyed (in conflict).

C | ARIES. The Fulfilment of the Promised Deliverance.

The Third Book.

E | TAURUS. The Prophecy of the coming Judge of all the earth.

  F | GEMINI. The Redeemer's reign (Grace and Glory).

  F | CANCER. The Redeemer's possession (safe from all conflict).

E | LEO. The fulfilment of the promised Triumph.

Hence in CAPRICORNUS we must look for the *prophecy* of this Coming Sacrifice. As a matter of fact it did actually point out the time when the Sun of Righteousness should arise, and "the Light of the World" appear. For when this Promised Seed was born the Sun *was actually in this sign of Capricornus!* "The fulness of time was come," and "God sent forth His Son TO REDEEM them that were under the Law" (Gal. iv. 4). The Sun was really amongst those very stars—*Al Gedi, the kid*, and *Deneb Al Gedi, the sacrifice cometh*—when this willing Sacrifice said, "Lo I come to do Thy will, O God." The nights were at their darkest and their longest when Jesus was born. The days began immediately to lengthen when He, "the true light," had come into the world.*

Astronomers confess that the perverted legends of the Greeks give but "a lame account" of this sign, "and it offers no illustration of *its ancient origin*."

Its ancient origin reveals a prophetic knowledge, which only He possessed who knew that in "the fulness of time" He would send forth His Son.

We now come to the three constellations which give us three pictures setting forth the death of this Sacrifice and of His living again.

§ 1. SAGITTA (The Arrow).

The Arrow of God sent forth.

It is not the Arrow of Sagittarius, for that has not left his bow. That arrow is for the enemies of God. This is for the Son of God. It was of this that He spoke when He said, in Ps. xxxviii. 2:

> "Thine arrows stick fast in me,
> And Thy hand presseth me sore."

He was "stricken, smitten of God, and afflicted, He was wounded for our transgressions" (Isa. liii. 4, 5). He was "pierced," when He could say with Job, "The arrows of the Almighty are within me" (vi. 4).

---

* When we come to the last chapter of this book we shall see that the Sun was in the sign of the other sacrificial animal, ARIES, at the very hour of the Crucifixion. And ARIES sets before us the victory of "the Lamb that was slain."

Here the arrow is pictured to us in mid-heaven, alone, as having been shot forth by an invisible hand. It is seen in its flight through the heavens. It is the arrow of God, showing that Redemption is all of God. It was "the will of God" which Jesus came to do. Not a mere work of mercy for miserable sinners, but a work ordained in eternity past, for the glory of God in eternity future.

This is the record of the Word, and this is what is pictured for us here. The work which the arrow accomplishes is seen in the dying Goat, and in the falling Eagle.

There are many other stars in the heavens in a straighter line, which would better serve for an arrow. Why are these stars chosen? Why is the arrow placed here? What explanation can be given, except that the Revelation in the stars and in the Book are both from the inspiration of the same Spirit?

There are about 18 stars, of which four are of the 4th magnitude. Only γ and δ are in the same line, while the shaft passes between α and β.

The Hebrew name is *Sham*, *destroying*, or *desolate*.

§ 2. AQUILA (The Eagle).

The Smitten One Falling.

Here we have an additional picture of the effect of this arrow, in the pierced, wounded, and falling Eagle, gasping in its dying struggle. And that

XVII. SAGITTA. (the arrow) AQUILA. (the Eagle) DELPHINUS. (the Dolphin)

pierced, wounded, and dying Saviour whom it represents, after saying, in Ps. xxxviii. 2, "Thine arrows stick fast in Me," added, in verse 10:

> "My heart panteth, My strength faileth Me,
> As for the light of Mine eyes it is gone from Me."
> (See also Zech. xiii. 6.)

The names of the stars, all of them, bear out this representation. The constellation contains 74 stars. The brightest of them, α (in the Eagle's neck), is a notable star of the 1st magnitude, called *Al Tair* (Arabic), *the wounding*. The star β (in the throat) is called *Al Shain* (Arabic), *the bright*, from a Hebrew root meaning *scarlet coloured*, as in Josh. ii. 18. The star γ (in the back) is called *Tarared*, *wounded*, or *torn*. δ (in the lower wing) is named *Alcair*, which means *the piercing*, and ε (in the tail), *Al Okal*, has the significant meaning *wounded in the heel*.

How can the united testimony of these names be explained except by acknowledging a Divine origin? even that of Him who afterwards foretold of the bruising of the Virgin's Son in the written Word; yea, of Him "who telleth the number of the stars and giveth them all their names."

§ 3. DELPHINUS (The Dolphin).

The Dead One Rising again.

This is a bright cluster of 18 stars, five of which are of the 3rd magnitude. It is easily distinguished by the four brightest, which are in the head.

It is always figured as a fish full of life, and always with the head upwards, just as the eagle is always with the head downwards. The great peculiar characteristic of the dolphin is its rising up, leaping, and springing out of the sea.

When we compare this with the dying goat and falling eagle, what conclusion can we come to but that we have here the filling in of the picture, and the completion of the whole truth set forth in Capricornus?

Jesus "died and rose again." Apart from His resurrection His death is without result. In His conflict with the enemy it is only His coming again in glory which is shown forth. But here, in connection with His people, with the multitudes of His redeemed, Resurrection is the great and important truth. He is "the first-fruits of them that slept"; then He, too, is here represented as a fish. He who went down into the waters of death for His people; He who could say "All thy waves and thy billows are gone over me" (Ps. xlii. 7), He it is who rises up again from the dead, having died on account of the sins of His redeemed, and risen again on account of their justification (Rom. iv. 25).

This is the picture here. In the Persian planisphere there seems to be a fish and a stream of water. The Egyptian has a vessel pouring out water.

The ancient names connected with this constellation are *Dalaph* (Hebrew), *pouring out of water;*

Dalaph (Arabic), *coming quickly; Scalooin* (Arabic), *swift (as the flow of water); Rotaneb* or *Rotaneu* (Syriac and Chaldee), *swiftly running*.

Thus, in this first chapter of the Second Book we see the great truth of Revelation set forth; and we learn how the great Blessings of Redemption were procured. This truth cannot be more eloquently or powerfully presented than in the language of Dr. Seiss:—

"This strange goat-fish, dying in its head, but living in its afterpart—falling as an eagle pierced and wounded by the arrow of death, but springing up from the dark waves with the matchless vigour and beauty of the dolphin—sinking under sin's condemnation, but rising again as sin's conqueror—developing new life out of death, and heralding a new springtime out of December's long drear nights—was framed by no blind chance of man. The story which it tells is the old, old story on which hangs the only availing hope that ever came, or ever can come, to Adam's race. To what it signifies we are for ever shut up as the only saving faith. In that dying Seed of the woman we must see our sin-bearer and the atonement for our guilt, or die ourselves unpardoned and unsanctified. Through His death and bloodshedding we must find our life, or the true life, which alone is life, we never can have."

"Complete atonement Thou hast made,
And to the utmost farthing paid
  Whate'er Thy people owed:
Nor can His wrath on me take place,
If sheltered in His righteousness,
  And sprinkled with the blood.

If my discharge Thou hast procured,
And freely in my room endured
  The whole of wrath divine,
Payment God cannot twice demand,
First at my bleeding Surety's hand,
  And then again at mine.

F 2

Turn, then, my soul, unto Thy rest;
The merits of Thy great High Priest
Have bought thy liberty;
Trust in His efficacious blood,
Nor fear thy banishment from God,
Since Jesus died for thee."

CHAPTER II.

THE SIGN AQUARIUS (The Water Bearer).

Their Blessings Ensured, or the Living Waters of Blessing Poured Forth for the Redeemed.

THE Atonement being made, the blessings have been procured, and now they can be bestowed and poured forth upon the Redeemed. This is the truth, whether we think of Abel's lamb, of patriarchal sacrifices, the offerings under the Law, or of that great Sacrifice of which they all testified. They all with one voice tell us that atonement made is the only foundation of blessing.

This was pictured and foreshown in the heavens from the beginning, by a man pouring forth water from an urn which seems to have an inexhaustible supply, and which flows forth downwards into the mouth of a fish, which receives it and drinks it all up.

In the ancient Zodiac of Denderah it is the same idea, though the man holds two urns, and the fish below seems to have come out of the urn. The

XVIII.
AQUARIUS.
(the Water Bearer)
& PISCIS AUSTRALIS
(the Southern Fish)
DELPHINUS
PISCES
ECLIPTIC
CETUS
CAPRICORNUS
SAGITTARIUS
FLUVIUS AQUARII
Fom al Haut
Scheat
Al Gedi
Deneb Al Gedi

[illegible]

[illegible] is called [illegible]

In some nations, I have [illegible] appear [illegible]

[illegible] the same.

There are [illegible] of which are of [illegible]

The star [illegible] shoulder is called [illegible] the record of the pitcher.

The star [illegible] is called [illegible]

The [illegible] in the lower part of the [illegible] which means [illegible]

[illegible] which [illegible] simply [illegible]

Aquarius is [illegible] the same meaning [illegible]

Can [illegible] what is this [illegible] the Greeks [illegible] of [illegible] the woman of Samaria [illegible] that living water which He alone can give. The [illegible] the same about [illegible] and another [illegible] not possible to reconcile the man is called *Hupei Tirion*, which means *the place of him coming down* or *poured forth.*

In some eastern Zodiacs the Urn alone appears.

This agrees with its other names—Hebrew, *Deli, the water-urn*, or *bucket* (as in Num. xxiv. 7); the Arabic *Delu* is the same.

There are 108 stars in this Sign, four of which are of the 3rd magnitude. Their names, as far as they have come down to us, are significant.

The star α (in the right shoulder) is called *Sa'ad al Melik*, which means *the record of the pouring forth.*

The star β (in the other shoulder) is called *Saad al Sund, who goeth and returneth*, or *the pourer out.*

The bright star δ (in the lower part of the right leg) is well-known to-day by its Hebrew name *Scheat*, which means *who goeth and returneth.*

The bright star in the urn has an Egyptian name—*Mon* or *Meon*, which means simply *an urn.*

Aquarius is the modern Latin name by which the sign is known. It has the same meaning, *the pourer forth of water.*

Can we doubt what is the interpretation of this sign? The Greeks, not knowing Him of whom it testified, were, like the woman of Samaria, destitute of that living water which He alone can give. They therefore invented some story about *Deucalion*, the son of Prometheus; and another, saying he is *Ganymede*, Jove's cup-bearer! But, as an astronomer says, "We must account otherwise for the origin of this name; for it is not possible to reconcile the symbols of the eleventh* sign with Grecian mythology." No! we must go further back than that, and not cramp our vision, and distort the Scriptures, by confining our thoughts to "the Church." The Church is nowhere seen in these Signs, as she is nowhere revealed in the Old Testament. This we shall enlarge on when we come to the sign Pisces. Meanwhile we must read the witness of the stars as if there had been no Church!

Christ is first. Yea, He is all in all. The Scriptures testify of Him; and the very stars in this Sign tell of His going away and His coming again. These prophetic signs have to do with Him, with the Atonement He wrought, with the conflict He endured, with the blessings He secured, with the victory He shall win, and the triumph He shall have. For it is written:

> "He shall pour the water out of His buckets,
> And His seed shall be in many waters,
> And His king shall be higher than Agag,
> And His kingdom shall be exalted."
>
> (Num. xxiv. 7.)

It tells of that glorious day when

> "A King shall reign in righteousness;
> And princes shall rule in judgment;
> And a MAN shall be as an hiding place
> from the wind,
> And a covert from the tempest;
> As RIVERS of WATER in a dry place."
>
> (Isa. xxxii. 1, 2).

---

* The eleventh, because everyone begins to reckon from ARIES, and not as we have done from VIRGO, as shown by the riddle of the Sphinx. See page 20.

It speaks of that glorious time when Israel shall be restored, and their "eyes shall see the King in His beauty"; when the peace of Zion shall be no more disturbed, "but there the glorious LORD will be unto us a place of broad rivers and streams" (Isa. xxxiii. 17, 20, 21). Then

> "The wilderness and the solitary place shall be glad
> for them;
> And the desert shall rejoice, and blossom as the rose,
> For in the wilderness shall waters break out,
> And streams in the desert." (Isa. xxxv. 1, 6.)

> "I will open rivers in high places,
> And fountains in the midst of the valleys;
> I will make the wilderness a pool of water,
> And the dry land springs of water." (Isa. xli. 18.)

> "Fear not, O Jacob, My servant;
> And thou, Jesurun, whom I have chosen,
> For I will POUR WATER upon him that is thirsty,
> And floods upon the dry ground;
> I will pour My Spirit upon thy seed,
> And My blessing upon thy offspring.
> Thus saith the LORD the King of Israel,
> And his Redeemer the LORD of hosts."
> (Isa. xliv. 2, 3, 6.)

This is the meaning of the Sign. The MAN Christ Jesus, who was humbled in death will yet be seen to be the pourer forth of every blessing. *Physically* pouring forth literal waters, removing the curse, and turning this world into a paradise:

> "Making her wilderness like Eden,
> And her desert like the garden of the LORD." (Isa. li. 3.)

And *morally* pouring forth His Spirit in such abundance as to fill the whole earth with peace, and blessing, and glory, "as the waters cover the sea."

Upon Israel restored He will pour out His blessing. They will be sprinkled with clean water, and possess a new heart and a new spirit (Ezek. xxxvi. 24–28; Joel ii. 28–32).

Such are some of the Scriptures which tell of this glorious Water-pourer. We need not rob Christ of His glory, or Israel of her blessing, in order to see in all this Pentecost or the Church. These are quite independent of the great line of prophetic truth. They are parenthetical, and distinct, and true, quite apart from the glorious prophecies of Israel's scattering and gathering. The physical marvels referred to in the texts above can never be satisfied or exhausted by any spiritual fulfilment. We may make an *application* of them as far as is consistent with the teaching of the epistles; but the *interpretation* of them belongs to the Person of Christ, and the nation of Israel. That interpretation is pictured for us in the Sign, and in its three constellations.

§ 1. PISCIS AUSTRALIS (The Southern Fish).

The Blessings Bestowed.

This first constellation is one of high antiquity,* and its brilliant star of the first magnitude was a subject of great study by the Egyptians and Ethiopians. It is named in Arabic *Fom al Haut*, *the mouth of the fish.* There are 22 other stars.

* And in great contrast with several modern ones near it, *e.g.*, the Balloon, the Sculptor's Apparatus, the Microscope, Euclid's Square, the Telescope, etc., etc.

XIX. PEGASUS (the Winged Horse)

The constellation is inseparable from Aquarius, in connection with which we have shown it in Plate XVIII. In the Denderah Zodiac it is called *Aar, a stream.*

It sets forth the simple truth that the blessings procured by the MAN—the coming Seed of the woman, will be surely bestowed and received by those for whom they are intended. There will be no failure in their communication, or in their reception. What has been purchased shall be secured and possessed.

§ 2. PEGASUS (The Winged Horse).

The Blessings Quickly Coming.

Not only shall they be received, but they shall be brought near. They will not have to be fetched, but they will be caused to come to those for whom they are procured, and will yet be *brought* by Him who has procured them.

In the Denderah Zodiac there are two characters immediately below the horse, *Pe* and *ka. Peka* or *Pega*, is in Hebrew *the chief*, and *Sus* is *horse.* So that the very word (*Pegasus*) has come down to us and has been preserved through all the languages.

The names of the stars in this constellation declare to us its meaning. There are 89 altogether; one of the 1st magnitude, two of the 2nd, three of the 3rd, nine of the 4th, etc. And, as astronomers testify, "they render Pegasus peculiarly remarkable."

The brightest, α (on the neck of the horse at the junction of the wing), comes down to us with the ancient Hebrew name of *Markab*, which means *returning from afar*. The star β (in the near shoulder) is called *Scheat*, i.e., *who goeth and returneth*. The star γ (at the tip of the wing) bears an Arabic name—*Al Genib*, *who carries*. The star ϵ (in the nostril) is called *Enif* (Arabic), *the water*. The star η (in the near leg) is called *Matar* (Arabic), *who causes to overflow*.

These names show us that we have to do with no mere horse. A winged horse is unknown to nature. It must therefore be used as a figure; and it can be a figure only of a person, even of Him who is "*the Branch*," as the star *Enif* shows, who said, "If I go away I will come again," as the star *Scheat* testifies.

He who procured these blessings for the redeemed by His Atonement, is quickly coming to bring them; and is soon returning to pour them forth upon a groaning creation. This is the lesson of Pegasus.

"Come, blessed Lord, bid every shore
  And answering island sing
The praises of Thy royal Name,
  And own Thee as their King.

Lord, Lord! Thy fair creation groans—
  The earth, the air, the sea—
In unison with all our hearts,
  And calls aloud for Thee.

Thine was the Cross with all its fruits
  Of grace and peace divine:
Be Thine the Crown of glory now,
  The palm of victory Thine."

XX.
CYGNUS
(the Swan)
DRACO
Etamin
Ras Taban
LYRA
Vega
HERCULES
Var 1686
Al Birrio
Deneb
Sadr
Al Deramin
Azelfafage
CEPHEUS
CASSIOPEIA
PEGASUS

§ 3. CYGNUS (The Swan).

The Blesser surely Returning.

This constellation repeats, emphasises, and affirms this glorious truth. It has to do with the Great Blesser and His speedy return, as is testified by all the ancient names connected with it.

In the Denderah Zodiac it is named *Tes-ark*, which means *this from afar*.

It is a most brilliant and gorgeous asterism of 81 stars; one of the 1st or 2nd, six of the 3rd, twelve of the 4th magnitude, etc. It contains variable stars, five double stars, and one quadruple. The star marked "61 Cygni" is known as one of the most wonderful in the whole heavens. It consists of two stars which revolve about each other, and yet have a progressive motion common to each!

This mighty bird is not falling dead, like Aquila, but it is flying swiftly in mid-heaven. It is coming to the earth, for it is not so much a bird of the air, but a bird peculiarly belonging to both the earth and the waters.

Its brightest star, α (between the body and the tail), is called *Deneb* (like another in Capricornus), and means *the judge*. It is also called *Adige*, *flying swiftly*, and thus at once it is connected with Him who cometh to judge the earth in righteousness.

The star β (in the beak) is named *Al Bireo* (Arabic), *flying quickly*.

The star γ (in the body) is called *Sadr* (Hebrew), *who returns as in a circle*.

The two stars in the tail, now marked in the maps as π 1 and π 2, are named *Azel, who goes and returns quickly;* and *Fafage, gloriously shining forth.*

The teaching, then, of the whole sign of AQUARIUS is clear and complete. The names of the stars explain the constellations, and the names of the constellations explain the sign, so that we are left in no doubt.

By His atoning death (as set forth in CAPRICORNUS) He has purchased and procured unspeakable blessings for His redeemed. This sign (AQUARIUS) tells of those blessings being poured forth, and of the speedy return of Him who is to bring "rivers of blessing," and to fill this earth with blessing and glory "as the waters cover the sea."

"Then take, LORD, thy kingdom, and come in Thy glory;
Make the scene of Thy sorrows the place of Thy throne,
Complete all the blessing which ages in story
Have told of the triumphs so justly Thine own."

CHAPTER III.

THE SIGN PISCES (The Fishes).

The Blessings of the Redeemed in abeyance.

IN this third chapter of the Second Book we come to the results of the Redeemer's work enjoyed, but in connection with conflict, as is seen in the last

XXI
PISCES the Fish and THE BAND

of the three sections (the constellation of *Andromeda, the chained woman*), which leads up to the last chapter of the book, and ends it in triumph over every enemy.

The Sign is pictured as two large fishes bound together by a *Band*, the ends of which are fastened separately to their tails. One fish is represented with its head pointing upwards towards the North Polar Star, the other is shown at right angles, swimming along the line of the ecliptic, or path of the sun.

The ancient Egyptian name, as shown on the Denderah Zodiac, is *Pi-cot Orion*, or *Pisces Hori*, which means *the fishes of Him that cometh.*

The Hebrew name is *Dagim, the Fishes*, which is closely connected with *multitudes*, as in Gen. xlviii. 26, where Jacob blesses Joseph's sons, and says, "Let them grow into a multitude in the midst of the earth." The margin says, "Let them grow *as fishes do increase*." It refers to the fulfilment of Gen. i. 28, "Be fruitful and multiply." The *multitude* of Abraham's seed is prominent in the pronouncement of the blessings, where God compared his future posterity to the stars of the sky, and the sand upon the sea shore. "A very great multitude of fish," as in Ezek. xlvii. 9.

The Syriac name is *Nuno, the fish, lengthened out (as in posterity).*

The sign, then, speaks of the multitudes who should enjoy the blessings of the Redeemer's work.

And here we must maintain that "the Church," which is "the Body of Christ," was a subject that was never revealed to man until it was made known to the Apostle Paul by a special revelation. The Holy Spirit declares (Rom. xvi. 25) that it "was kept secret since the world began." In Eph. iii. 9 he declares that it "from the beginning of the world hath been hid in God"; and in Col. i. 26, that it "hath been hid from ages and from generations, but now is made manifest to His saints." In each scripture which speaks of it as "now made manifest," or "now made known," it is distinctly stated that it was "a mystery," *i.e.*, a *secret*, and had, up to that moment, been hidden from mankind, hidden "in God." How, then, we ask, can "the Church," which was *a subsequent* revelation, be read into the previous prophecies, whether written in the Old Testament Scriptures, or made known in the Heavens? If the Church was revealed in prophecy, then it could not have been said to be hidden or kept secret. If the *first* revelation of it was made known to Paul, as he distinctly affirms it was, then it could not have been revealed before. Unless we see this very clearly, we cannot "rightly divide the word of truth" (2 Tim. ii. 15). And if we do not rightly divide the word of truth, in its subjects, and times, and dispensations, we must inevitably be landed in confusion and darkness, interpreting of the Church, scriptures which belong only to Israel.

The Church, or Body of Christ, is totally distinct from every class of persons who are made the subject of prophecy. Not that the Church of God was an after-thought. No, it was a Divine secret, kept as only God Himself could keep it. The Bible therefore would have been complete (so far as the Old Testament prophecies are concerned) if the Epistles (which belong only to the Church) were taken out. The Old Testament would then give us the kingdom prophesied; the Gospels and Acts, the King and the kingdom offered and rejected; then the Apocalypse would follow, showing how that promised kingdom will yet be set up with Divine judgment, power, and glory.

If these Signs and these star-pictures be the results of inspired patriarchs, then this Sign of Pisces can refer to "His seed," prophesied of in Isa. liii.: "He shall see His seed." It must refer to

> "The nation whose God is the Lord,
> And the people whom He hath chosen for His own inheritance." (Ps. xxxiii. 12.)

> "Such as be blessed of Him shall inherit the earth." (Ps. xxxvii. 22.)

> "The Lord shall increase you more and more,
> You and your children,
> Ye are blessed of the Lord." (Ps. cxv. 14, 15.)

> "Their seed shall be known among the Gentiles,
> And their offspring among the people;
> All that see them shall acknowledge them,
> That they are the seed which the Lord hath blessed." (Isa. lxi. 9.)

> "They are the seed of the blessed of the Lord,
> And their offspring with them." (Isa. lxv. 23.)

The prophecy of this Sign was afterwards written in the words of Isa. xxvi. 15—the song which shall yet be sung in the land of Judah:

> "Thou hast increased the nation, O LORD,
> Thou hast increased the nation."

And in Isa. ix. 3 (R.V.), speaking of the glorious time when the government shall be upon the shoulder of the coming King:

> "Thou hast multiplied the nation,
> Thou hast increased their joy."

Of that longed-for day Jeremiah sings (xxx. 19):

> "I will multiply them
> And they shall not be few;
> I will also glorify them,
> And they shall not be small."

Ezekiel also is inspired to say:

> "I will multiply men upon you,
> All the house of Israel, even all of it:
> And the cities shall be inhabited,
> And the wastes shall be builded;
> And I will multiply upon you man and beast;
> And they shall increase and bring fruit."
> (Ezek. xxxvi. 10, 11.)

> "Moreover I will make a covenant of peace with them;
> It shall be an everlasting covenant with them!
> And I will place them, and multiply them,
> And will set My sanctuary in the midst of them for evermore." (Ezek. xxxvii. 26.)

Indeed, this Sign of PISCES has always been interpreted of Israel. Both Jews and Gentiles have agreed in this. ABARBANEL, a Jewish commentator, writing on Daniel, affirms that the Sign PISCES always refers to the people of Israel. He gives five reasons for this belief, and also affirms that a conjunction of the planets Jupiter and Saturn always betokens a crisis in the affairs of Israel. Because such a conjunction took place in his day (about 1480 A.D.) he looked for the coming of Messiah.*

Certain it is, that when the sun is in PISCES all the constellations which are considered *noxious*, are seen above the horizon. What is true in astronomical observation is true also in historical fact. When God's favour is shown to Israel, "the Jew's enemy" puts forth his malignant powers. When they increased and multiplied in Egypt, he endeavoured to compass the destruction of the nation by destroying the male children; but their great Deliverer remembered His covenant, defeated the designs of the enemy, and brought the counsel of the heathen to nought. So it was in Persia; and so it will yet be again when the hour of Israel's final deliverance has come.

There can be no doubt that we have in this Sign the foreshowing of the multiplication and blessing of the children of promise, and a token of their coming deliverance from all the power of the enemy.

But why *two* fishes? and why is one horizontal and the other perpendicular? The answer is, that not only in Israel, but in the seed of Seth and Shem there were always those who looked for a heavenly portion, and were "partakers of a heavenly calling." In Heb. xi. we are distinctly told that Abraham

* How inconsistent when there were three such conjunctions in one year, all in the same sign of PISCES, immediately preceding the birth of the woman's Seed; and in addition to this the new star which had been foretold. See under *Coma*, pages 36, 37, 38.

"looked for a city which hath foundations, whose builder and maker is God" (*v.* 10). They were "strangers and pilgrims on the earth" (*v.* 13). *Strangers* are those without a home, and *pilgrims* are those who are journeying home: "they seek a country" (*v.* 14). They desired "a better country, that is, an HEAVENLY: wherefore God is not ashamed* to be called their God; for He hath prepared for them a city" (*v.* 16). It is clear, therefore, that what are called the "Old Testament Saints" were "partakers of THE HEAVENLY CALLING" (Heb. iii. 1), which included a heavenly portion and a heavenly home; and all through the ages there have been "partakers of the heavenly calling." This is quite distinct from the calling of the Church, which is from both Jews and Gentiles to form "one body," a "new man" in Christ (Eph. ii. 15). It must be distinct, for it is expressly stated at the end of that chapter (Heb. xi. 40) that God has "PROVIDED (marg. *foreseen*) SOME BETTER THING FOR US." How can this be a "better thing," if it is the *same thing?* There must be two separate things if one is "better" than the other! Our calling in Christ is the "better thing." The Old Testament saints had, and will have, *a good thing.* They will have a heavenly blessing, and a heavenly portion, for God has "prepared for them a city," and we see that prepared city, even "the holy city, new Jerusalem, coming down from God out of HEAVEN,

* The figure of *Tapeinosis,* which calls our attention to that fact that He was delighted thus to be called.

prepared as a bride adorned for her husband" (Rev. xxi. 2). This is the "heavenly" portion of the Old Testament saints, the Bride of Christ. The Church will have a still "better" portion, for "they without us should not be made perfect" (Heb. xi. 40).

The fish, shooting upwards to the Polar Star, exquisitely pictures this "heavenly calling"; while the other fish, keeping on the horizontal line, answers to those who were content with an earthly portion.

But both alike were divinely called, and chosen, and upheld. The names of two of the stars in the sign (not identified) are *Okda* (Hebrew), *the united*, and *Al Samaca* (Arabic), *the upheld*.* These again speak of the redeemed seed, of whom, and to whom, Jehovah speaks in that coming day of glory in Isa. xli. 8–10 (R.V.):—

"But thou, Israel, My servant,
Jacob, whom I have chosen,
The seed of Abraham My friend;
Thou whom I have taken hold of from the ends
of the earth,
And called thee from the corners thereof,
And said unto thee, Thou art My servant:
I have chosen thee, and not cast thee away;
Fear thou not, for I am with thee;
Be not dismayed, for I am thy God!
I will strengthen thee;
Yea, I will help thee;
Yea, I will UPHOLD thee with the right hand
of My righteousness."

This is the teaching of the Sign; and the first constellation takes up this thought and emphasises it.

---

* There are 113 stars in this sign, none of any great importance; only one of the 3rd magnitude, five of the 4th, etc.

G 2

§ 1. THE BAND.

The Redeemed Bound, but binding their Enemy.

The band that *unites* these two fishes has always formed a separate constellation. It is shown in Plate XXI. The Arabian poems of ANTARAH frequently mention it as distinct from the Sign with which it is so closely connected. ANTARAH was an Arabian poet of the sixth century.

Its ancient Egyptian name was *U-or*, which means *He cometh.* Its Arabic name is *Al Risha*, *the band*, or *bridle.*

It speaks of the Coming One, not in His relation to Himself, or to His enemies, but in His relation to *the Redeemed.* It speaks of Him who says:

> "I drew them with cords of a man,
> With bands of love;
> And I was to them as they that take off the yoke
> on their jaws." (Hosea xi. 4, R.V.)

But it speaks also of His unloosing the bands with which they have been so long bound.

In the picture these fishes are bound. One end of the *band* is fastened securely round the tail of one fish, and it is the same with the other. Moreover, this *band* is fastened to the neck of *Cetus*, the sea monster, while immediately above is seen a woman chained as a captive. These both tell the same story, and, indeed, all are required to set forth the whole truth. The fishes are *bound* to *Cetus;* the woman (*Andromeda*) is chained; but the Deliverer of

XXII ANDROMEDA
(the Chained Woman)
PEGASUS
Al Phiratz
PISCES
Mirach
CASSIOPEIA
Al Mach or Al Amak
PERSEUS
Algol
Al Genib
CAPUT MEDUSÆ

both is near. Cepheus, the Crowned King, the Redeemer, "the Breaker," the Branch, is seen coming quickly for the deliverance of His redeemed. These are the three constellations of this sign, and all three are required to set forth the story.

Israel now is bound. The great enemy still oppresses, but deliverance is sure. ARIES, *the Ram*, is seen with his paws on this band, as though about to loosen the bands and set the captives free, and to fast bind their great oppressor.

§ 2. ANDROMEDA (The Chained Woman).

The Redeemed in their Bondage and Affliction.

This is a peculiar picture to set in the heavens. A woman with chains fastened to her feet and arms, in misery and trouble; and bound, helpless, to the sky. Yet this is the ancient foreshowing of the truth.

In the Denderah Zodiac her name is *Set*, which means *set, set up as a queen*. In Hebrew it is *Sirra, the chained*, and *Persea, the stretched out*.

There are 63 stars in this constellation, three of which are of the 2nd magnitude, two of the 3rd, twelve of the 4th, etc.

The brightest star, α (in the head), is called *Al Phiratz* (Arabic), *the broken down*. The star β (in the body) is called *Mirach* (Hebrew), *the weak*. The star γ (in the left foot) is called *Al Maach*, or *Al Amak* (Arabic), *struck down*.

The names of other stars, not identified, are *Adhil, the afflicted; Mizar, the weak; Al Mara* (Arabic), *the afflicted.* ARATUS speaks of *Desma*, which means *the bound*, and says—

> "Her feet point to her bridegroom
> *Perseus*, on whose shoulder they rest."

Thus, with one voice, the stars of *Andromeda* speak to us of the captive daughter of Zion. And her coming Deliverer thus addresses her:—

> "O thou afflicted, tossed with tempest, and not comforted,
> Behold, . . . in righteousness shalt thou be established:
> Thou shalt be far from oppression; for thou shalt not fear:
> And from terror; for it shall not come nigh thee."
>
> (Isa. liv. 11–14.)

> "Hear now this, thou afflicted. . . .
> Awake, awake; put on thy strength, O Zion;
> Put on thy beautiful garments, O Jerusalem. . . .
> Shake thyself from the dust;
> Arise, and sit down, O Jerusalem:
> Loose thyself from the bands of thy neck, O captive daughter of Zion.
> For thus saith the LORD, Ye have sold yourselves for nought;
> And ye shall be redeemed without money."
>
> (Isa. li. 21—lii. 3.)

"The virgin daughter of My people is broken with a great breach, with a very grievous blow" (Jer. xiv. 17).

The picture which sets forth her deliverance is reserved for the next chapter (or Sign), where it comes in its proper place and order. We are first shown her glorious Deliverer; for we never, in the heavens or in the Word, have a reference to the sufferings without an *immediate* reference to the glory.

Al Deramin
Al Phirk
Al Rai
Caph
CASSIOPEIA
DRACO
Al Ruccaba
Polar Star
XXIII
CEPHEUS
(the Crowned King)

§ 3. CEPHEUS (The King).

Their Redeemer Coming to Rule.

Here we have the presentation of a glorious king, crowned, and enthroned in the highest heaven, with a sceptre in his hand, and his foot planted on the very Polar Star itself.

His name in the Denderah Zodiac is *Pe-ku-hor*, which means *this one cometh to rule.*

The Greek name by which he is now known, *Cepheus*, is from the Hebrew, and means *the branch*, and is called by EURIPIDES *the king*.

An old Ethiopian name was *Hyk*, *a king*.

There are 35 stars, *viz.*, three of the 3rd magnitude, seven of the 4th, etc.

The brightest star, α (in the left shoulder), is called *Al Deramin*, which means *coming quickly*. The next, β (in the girdle), is named *Al Phirk* (Arabic), *the Redeemer*. The next, γ (in the left knee), is called *Al Rai*, which means *who bruises* or *breaks*.

It is impossible to mistake the truth which these names teach. The Greeks, though they had lost it, yet preserved a trace of it, even in their perversion of it; for they held that *Cepheus* was the father of *Andromeda*, and that *Perseus* was her husband.

Yes; this is the glorious King of Israel, the "King of kings, and Lord of lords." It is He who calls Israel His "son," and will yet manifest it to all the world.

In Jer. xxxi., after speaking of Israel's restoration, Jehovah says (*v.* 1):—

"At the same time, saith the LORD, will I be the God
of all the families of Israel,
And they shall be My people. . . .
For I am a father to Israel,
And Ephraim is My firstborn" (*v.* 9).

As He said to Moses: "Thus saith the LORD, Israel is my son, even my firstborn" (Exod. iv. 22).

Here is the foundation of Israel's blessing. True, it is now in abeyance, but "the LORD reigneth," and will in due time make good His Word, for

"The counsel of the LORD standeth for ever.
The thoughts of His heart to all generations."
(Ps. xxxiii. 11.)

This leads us up to the last chapter of the Second Book, which shows us the fulfilment of all the prophecies concerning the Redeemed and the sure foundation on which their great hope of glory is based.

CHAPTER IV.

THE SIGN ARIES (The Ram or Lamb).

The Blessings of the Redeemed Consummated and Enjoyed.

THIS Second Book began with *the Goat* dying in sacrifice, and it ends with the Lamb living again, "as it had been slain." The goat had the tail of a fish, indicating that his death was for a *multitude* of

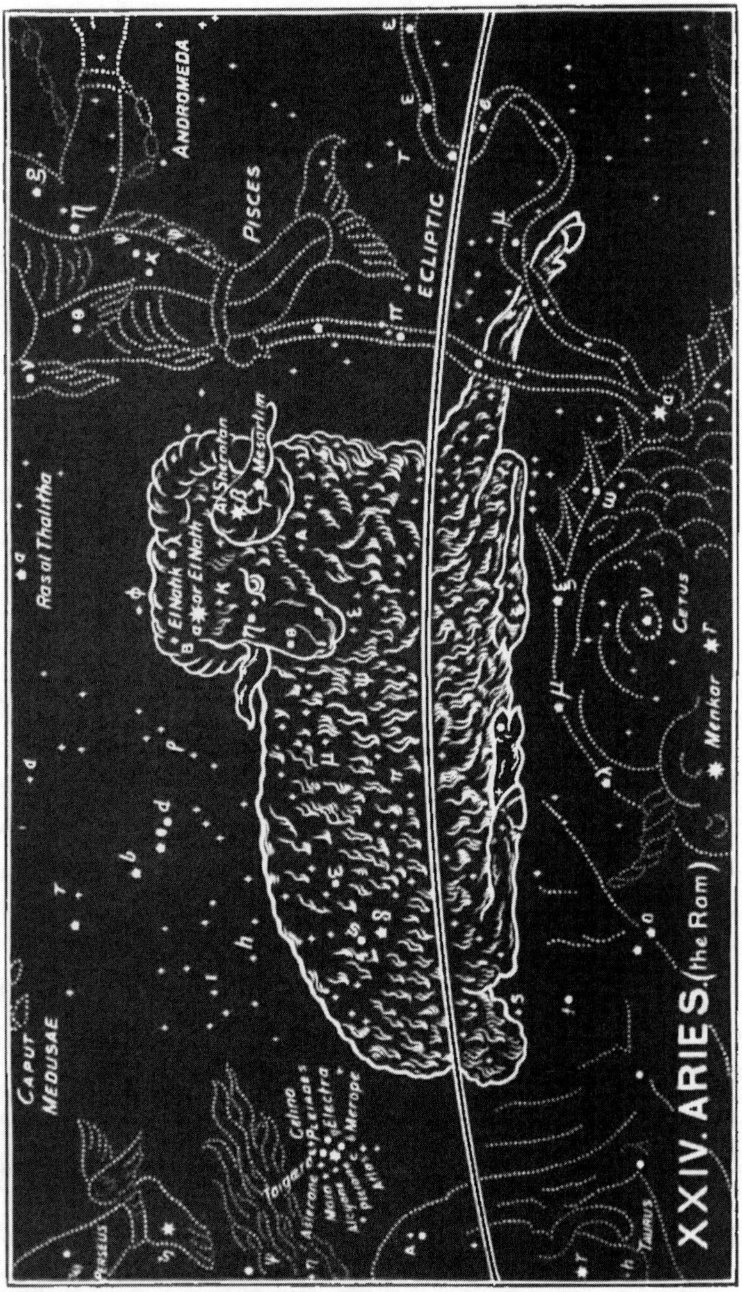

XXIV. ARIES (the Ram)

the redeemed. In the two middle Signs we have had these fishes presented to us in grace, and in their conflict. We come now to the last chapter of the book: and, as we have seen, like each of the other books, it ends up with victory and triumph. Here we are first shown the foundation on which that victory rests, namely, Atonement. Hence we are taken back and reminded of the "blood of the Lamb."

This is pictured by a ram, or lamb, full of vigour and life; not falling in death as *Capricornus* is.

In the Denderah Zodiac its name is *Tametouris Ammon*, which means *the reign*, *dominion*, or *government of Ammon*. The lamb's head is without horns, and is crowned with a circle.

The Hebrew name is *Taleh*, *the lamb*. The Arabic name is *Al Hamal*, *the sheep*, *gentle*, *merciful*. This name has been mistakenly given by some to the principal star, α. The Syriac name is *Amroo*, as in the Syriac New Testament in John i. 29: "Behold the Lamb of God which taketh away the sin of the world." The ancient Akkadian name was *Baraziggar*. *Bar* means *altar*, or *sacrifice*; and *ziggar* means *right making*; so that the full name would be *the sacrifice of righteousness*.

There are 66 stars in this sign, one being of the 2nd magnitude, two of the 4th, etc.

Its chief star, α (in the forehead), is named *El Nath*,* or *El Natik*, which means *wounded*, *slain*. The next, β (in the left horn), is called *Al Sheratan*,

* "El Nath" is used by Chaucer as the name of a spring star.

the bruised, the wounded. The next, γ (near to β), is called *Mesartim* (Hebrew), *the bound.*

How is it there is no conflicting voice? How is it that all the stars unite in one harmonious voice in testifying of the Lamb of God, slain, and bruised, but yet living for evermore, singing together, "Worthy is the Lamb that was slain to receive power and riches, and wisdom, and strength, and honour, and glory, and blessing" (Rev. v. 12)?

This rejoicing connected with the Lamb shines faintly through the heathen perversions and myths: for HERODOTUS tells us how the ancient Egyptians, once a year, when it opened by the entrance of the sun into ARIES,* slew a Ram, at the festival of Jupiter Ammon; branches were placed over the doors, the Ram was garlanded with wreaths of flowers and carried in procession. Now the sun entered ARIES on the 14th of the Jewish month Nisan, and *another lamb* was then ordered to be slain, even "the LORD's passover"—the type of that Lamb that should in the fulness of time be offered without blemish and without spot. Owing to the precession of the equinoxes, the sun, at the time of the Exodus, had receded into this sign of ARIES, which then marked the Spring Equinox. But by the time that the antitype—the Lamb of God, was slain, the sun had still further receded, and on the 14th of Nisan, in the year of the Crucifixion, stood at the very spot marked by the stars α, *El Nath*, *the pierced*, *the wounded* or *slain*,

* TAURUS then marked the Spring Equinox.

and β, *Al Sheratan, the bruised* or *wounded!* God so ordained "the times and seasons" that during that noon-day darkness the sun was seen near those stars which had spoken for so many centuries of this bruising of the woman's Seed—the Lamb of God.

Was this design? or was it chance? It is far easier to believe the former. It makes a smaller demand upon our faith; yes, we are compelled to believe that He who created the sun and the stars "for signs and for cycles," ordained also the times and the seasons, and it is He who tells us that "WHEN THE FULNESS OF TIME WAS COME, God sent forth His Son" (Gal. iv. 4), and that "in due time Christ died for the ungodly" (Rom. v. 6).

§ 1. CASSIOPEIA (The Enthroned Woman).

The Captive Delivered, and Preparing for her Husband, the Redeemer.

In the last chapter we saw the *woman bound;* here we see the same woman freed, delivered, and enthroned.

ULUGH BEY says its Arabic name is *El Seder*, which means *the freed.*

In the Denderah Zodiac her name is *Set*, which means *set, set up as Queen.* ALBUMAZER says this constellation was anciently called "*the daughter of splendour.*" This appears to be the meaning of the word *Cassiopeia, the enthroned, the beautiful.* The Arabic name is *Ruchba, the enthroned.* This is also the meaning of its Chaldee name, *Dat al cursa.*

There are 55 stars in this constellation, of which five are of the 3rd magnitude, five of the 4th, etc.

This beautiful constellation passes vertically over Great Britain every day, and is easily distinguished by its five brightest stars, forming an irregular **W.**

This brilliant constellation contains one binary star, a triple star, a double star, a quadruple star, and a large number of nebulæ.

In the year 1572 Tycho Brahe discovered in this constellation, and very near the star κ (under the arm of the chair), a new star, which shone more brightly than Venus. It was observed for nearly two years, and disappeared entirely in 1574.

The brightest star, α (in the left breast), is named *Schedir* (Hebrew), which means *the freed*. The next, β (in the top of the chair), likewise bears a Hebrew name—*Caph*, which means *the branch;* it is evidently given on account of the branch of victory which she bears in her hand.

She is indeed highly exalted, and making herself ready. Her hands, no longer bound, are engaged in this happy work. With her right hand she is arranging her robes, while with her left she is adorning her hair. She is seated upon the Arctic circle, and close by the side of *Cepheus*, the King.

This is "the Bride, the Lamb's wife, the heavenly city, the new Jerusalem," the "partakers of the heavenly calling."

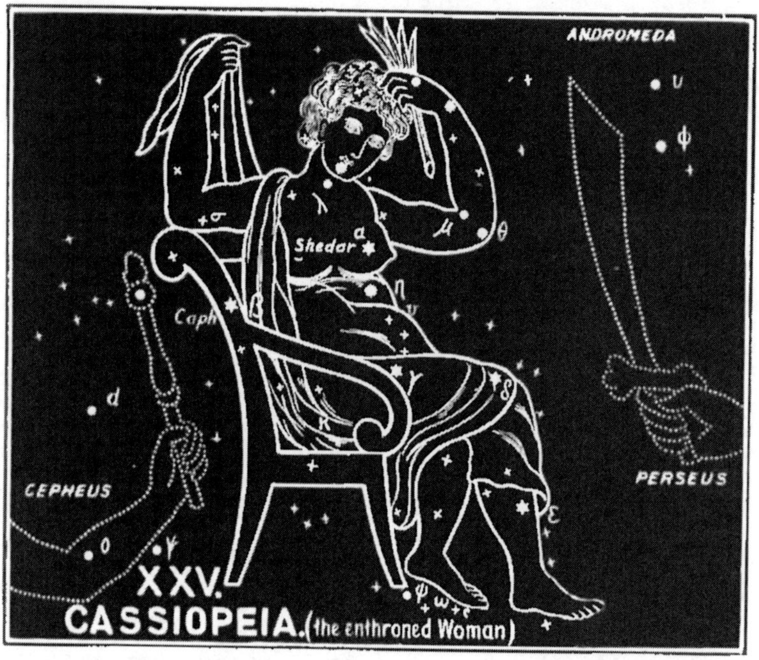

XXV.
CASSIOPEIA. (the enthroned Woman)

He who has redeemed her is "the Lamb that was slain," and He addresses her thus:

"Thy Maker is thine husband;
The LORD of Hosts is His name;
And the Holy One of Israel is thy Redeemer;
The God of the whole earth shall He be called.
For the LORD hath called thee as a woman forsaken and grieved in spirit,
Even a wife of youth when she is cast off, saith thy God.
For a small moment have I forsaken thee;
But with great mercies will I gather thee.
In overflowing wrath I hid my face from thee for a moment;
But with everlasting kindness will I have mercy on thee, saith the LORD thy Redeemer."

(Isa. liv. 5–8, R.V.)

"Thou shalt be a crown of beauty in the hand of the LORD,
And a royal diadem in the hand of thy God,
Thou shalt no more be termed Forsaken;
Neither shall thy land any more be termed Desolate;
But thou shalt be called Hephzi-bah (i.e., *my delight is in her*),
And thy land Beulah (i.e., *married*);
For the LORD delighteth in thee,
And thy land shall be married.
For as a young man marrieth a virgin,
So shall thy sons (*Heb.* thy Restorer) marry thee:
And as the bridegroom rejoiceth over the bride,
So shall thy God rejoice over thee."

(Isa. lxii. 3–5, R.V.)

"The LORD hath appeared of old (or from afar) unto me, *saying*,
Yea, I have loved thee with an everlasting love;
Therefore with lovingkindness have I drawn thee.
Again will I build thee, and thou shalt be built, O Virgin of Israel. . . .

He that scattered Israel will gather him,
And keep him as a shepherd doth his flock,
For the LORD hath ransomed Jacob,
And redeemed him from the hand of him that was stronger than he." (Jer. xxxi. 3–12, R.V.)

Can we close our eyes to the testimony of these scriptures—that Israel is the Bride of the Lamb? When we have all these, and more, why should we read "the Church" into these ancient prophecies, which was the subject of a long-subsequent revelation, merely because (in Eph. v. 25) Christ's love to His Church is *compared* to a husband's love for his wife? "Husbands, love your wives, even AS Christ also loved the Church." There is not a word here about the Church being His wife. On the contrary, it reveals the secret that the Church of Christ is to be the mystical "Body of Christ," *part of the Husband* in fact, "One new man" (Eph. ii. 15)! whereas restored Israel is to be the Bride of this "New Man," the Bride of Christ, the Lamb's wife! Blessed indeed it is to be united to Christ as a wife to a husband, but glorious beyond all description to be "one" with Christ Himself, part of His mystical Body.

If men had only realised the wondrous glory of this mystery, they would never have so *wrongly* divided the Word of Truth by *interpreting* Psalm xlv. of this Mystical Christ. If we "rightly divide" it, we see at once that this Psalm is in harmony with all the Old Testament scriptures, which must be interpreted alike, and can be interpreted only of Israel however they may be *applied*.

Having spoken of the Godhead and glory of this King (faintly and in part foreshown by *Cepheus*), the Holy Spirit goes on in the latter part of the Psalm to speak of the Bride—the Queen:

"At Thy right hand doth stand the Queen in gold of Ophir,
Hearken, O daughter, and consider, incline thine ear;
Forget also thine own people, and thy father's house;
So shall the King desire thy beauty;*
For He is thy Lord; and worship thou Him. . . .
The King's daughter within *the palace* is all glorious;
Her clothing is inwrought with gold,
She shall be led unto the King in broidered work;
The virgins her companions † that follow her shall be brought unto thee," etc. (Ps. xlv. 9–17, R.V.)

Then shall she sing her Magnificat:

"I will greatly rejoice in the LORD,
My soul shall be joyful in my God;
For He hath clothed me with the garments of salvation,
He hath covered me with the robe of righteousness,
As a bridegroom decketh *himself* with ornaments,
And as a bride adorneth *herself* with her jewels.
For as the earth bringeth forth her bud,
And as the garden causeth the things that are sown in it to spring forth;
So the Lord GOD [Adonai Jehovah] will cause righteousness and praise to spring forth before all the nations." (Isa. lxi. 10, 11.)

* "Thy beauty; for it was perfect through My comeliness, which I put upon thee (Jerusalem), saith the LORD" (Ezek. xvi. 14).

† Those who interpret the Queen here of the Church as the Bride, interpret the "Virgins" in Matt. xxv. of the Bride also. But how inconsistent! If the "Virgins" be the Church in Matt. xxv., then where is the Bride? If the Queen is the Bride (the Church) in Ps. xlv., then who are the "virgins her companions"? Both cannot be the correct interpretation. In fact, both are wrong, and hence the *confusion*. The Bride must be interpreted by the Old Testament scriptures, and the Prophecies which belong to Israel must not be robbed and given to the Church. They cannot be thus diverted without bringing confusion into the Scripture, and causing loss to our souls.

This, then, is the truth set forth by this enthroned woman. The blessing founded on Atonement, and the Redemption wrought by the Lamb that was slain, result in a glorious answer to Israel's prayer, "Turn our captivity, O Lord" (Ps. cxxvi. 4): when they that have "sown in tears shall reap in joy," and the Lord shall loosen her bonds, and place her enthroned by His side.

This, however, involves the destruction of her enemy, and this is what we see in the next section.

§ 2. CETUS (The Sea Monster).

The Great Enemy Bound.

When John sees the New Jerusalem, the Bride, the Lamb's wife (Rev. xxi. 10, 2), Satan has been bound already: for we read, a few verses before (xx. 1–3): "I saw an angel come down from heaven, having the key of the bottomless pit and a great chain in his hand. And he laid hold of the dragon, that Old Serpent, which is the Devil, and Satan, and bound him [*and kept him bound*] a thousand years, and cast him into the bottomless pit, and shut him up, and set a seal upon him, that he should deceive the nations no more, till the thousand years should be fulfilled."

This is what we see in the second section of this chapter—the second constellation in Aries.

The picture is that of a great Sea-monster, the largest of all the constellations. It is the natural

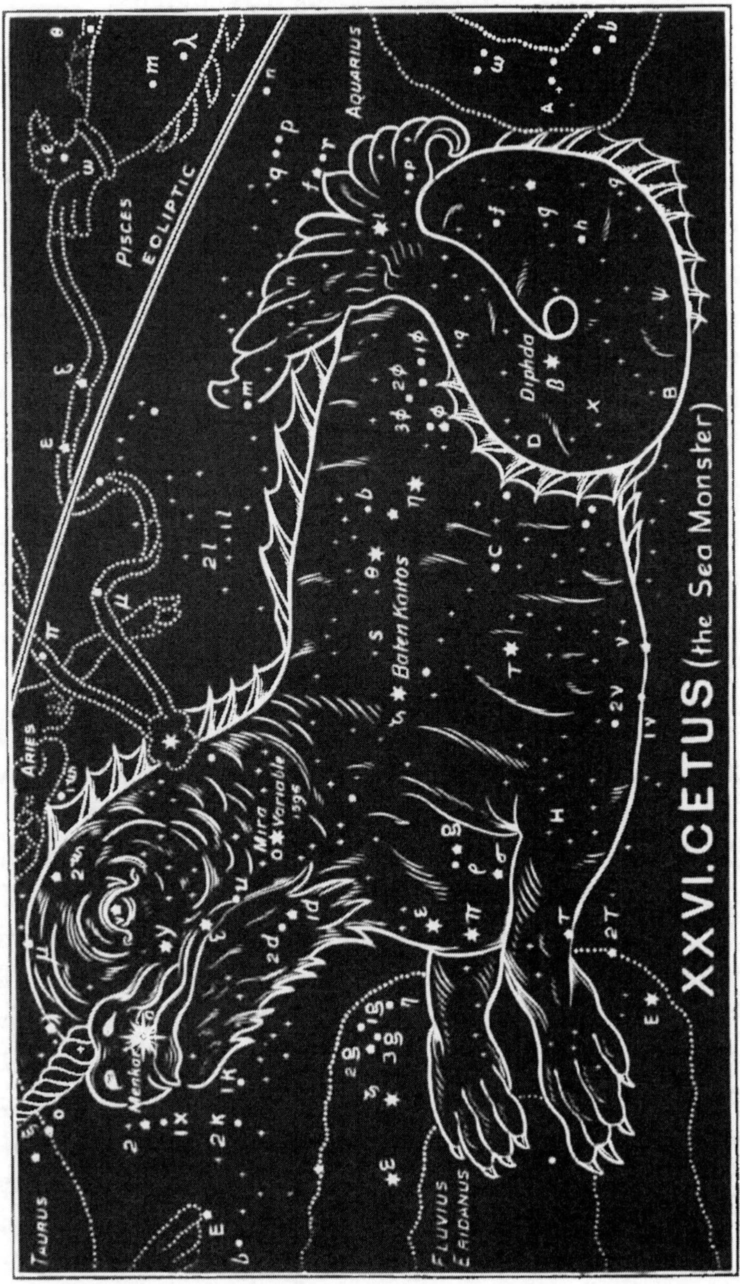

XXVI. CETUS (the Sea Monster)

enemy of fishes, hence it is placed here in connection with this last chapter, in which fishes are so prominent.

It is situated very low down among the constellations—far away towards the south or lower regions of the sky.

Its name in the Denderah Zodiac is *Knem*, which means *subdued*. It is pictured as a monstrous head, trodden under foot by the swine, the natural enemy of the serpent. The hawk also (another enemy of the serpent) is over this figure, crowned with a mortar, denoting *bruising*.

It consists of 97 stars, of which two are of the 2nd magnitude, eight of the 3rd, nine of the 4th, etc.

The names of the stars interpret for us infallibly the meaning of the picture.

The brightest star, α (in the upper mandible), is named *Menkar*, and means *the bound* or *chained enemy*. The next, β (in the tail), is called *Diphda*, or *Deneb Kaitos*, *overthrown*, or *thrust down*. The star o (in the neck) is named *Mira*, which means *THE REBEL*. Its name is ominous, for the star is one of the most remarkable. It is very bright, but it was not till 1596 that it was discovered to be *variable*. It disappears periodically *seven* times in *six* years! It continues at its brightest for fifteen days together. M. Bade says that during 334 days it shines with its greatest light, then it diminishes, till it entirely disappears for some time (to the naked eye). In fact, during that period it passes through several degrees of magnitude, both increasing and diminishing. Indeed its variableness is so great as to make it appear *unsteady!*

Here, then, is the picture of the Great Rebel as shown in the heavens. What is it, as written in the Word?

The Almighty asks man:—

"Canst thou draw out Leviathan with a fish hook?
Or press down his tongue with a cord?
Canst thou put a rope into his nose?
Or pierce his jaw through with a hook? . . . .
Shall not one be cast down even at the sight of him?
None is so fierce that he dare stir him up."
(Job xli. 1–10, R.V.)

But he whom man cannot bind can be bound by the Lamb, and He is seen with "the Band" that has bound the fishes, now in His hands, which he has fastened with a bright star to his neck, saying,—

"Behold, I have taken out of thine hand the cup of trembling,
Even the dregs of the cup of My fury;
Thou shalt no more drink it again,
But I will put it into the hand of them that afflict thee." (Isa. li. 22, 23.)

"Behold, the LORD cometh forth out of His place
To punish the inhabitants of the earth for their iniquity . . . .
In that day the LORD, with His sore, and great, and strong sword,
Shall punish Leviathan, the piercing serpent,
And Leviathan, the crooked serpent;
And He shall slay the dragon that is in the sea."
(Isa. xxvi. 21—xxvii. 1.)

"For God is my king of old,
Working salvation in the midst of the earth.
Thou didst divide (marg. *Heb., break*) the sea by Thy strength,
Thou brakest the heads of the dragons (R.V. marg., *sea monsters*) in the waters.
Thou brakest the heads of Leviathan in pieces."
(Ps. lxxiv. 12–14.)

And this Second Book closes by revealing to us this glorious "Breaker."

§ 3. PERSEUS. ("The Breaker.")

"The Breaker" delivering His Redeemed.

Here we have set before us a mighty man, called in the Hebrew *Peretz*, from which we have the Greek form *Perses*, or *Perseus* (Rom. xvi. 13). It is the same word which is used of Christ in Micah ii. 13. When He shall surely "gather the remnant of Israel" (*v.* 12), it is written—

"THE BREAKER is gone up before them . . . .
Their King is passed on before them,
And the LORD at the head of them."

This is what is pictured to us here. We see a glorious "Breaker" taking His place before His redeemed, breaking forth at their head, breaking down all barriers, and breaking the heads of Leviathan and all his hosts. In His right hand He has His "sore, and great, and strong sword" lifted up to smite and break down the enemy. He has wings on His feet, which tell us that He is coming very swiftly. In His left hand He carries the head of the enemy, whom he has slain.

H 2

In the Denderah Zodiac His name is *Kar Knem, he who fights and subdues.*

It is a beautiful constellation of 59 stars, two of which are of the 2nd magnitude, four of the 3rd, twelve of the 4th, etc.

Their names supply us with the key to the interpretation of the picture.

The star α (in the waist) is called *Mirfak, who helps.* The next, γ (in the right shoulder), is named *Al Genib,* which means *who carries away.* The bright star in the left foot is called *Athik, who breaks!*

In his left hand he carries a head, which, by perversion, the Greeks called the head of Medusa, being ignorant that its Hebrew root meant *the trodden under foot.** It is also called *Rosh Satan* (Hebrew), *the head of the adversary,* and *Al Oneh* (Arabic), *the subdued,* or *Al Ghoul, the evil spirit.*

The bright star, β (in this head), has come down to us with the name *Al Gol,* which means *rolling round.*

It is a most remarkable phenomenon that so many of these enemies should be characterised by variable stars! But this head of *Medusa,* like the neck of *Cetus,* has one. *Al Gol* is continually changing. In about 69 hours it changes from the 4th magnitude to the 2nd. During four hours of this period it gradually diminishes in brightness, which it recovers in the succeeding four hours; and in the remaining part of the time invariably preserves its greatest lustre. After

* See Job xxxix. 14, 15, where it is said, the ostrich "leaveth her eggs in the dust, forgetting that the foot may crush them, or that the wild beast may break them."

XXVII. PERSEUS (the Breaker)

the expiration of this time its brightness begins to decrease again. Fit emblem of our great enemy, who, "like *a roaring lion*, goeth about seeking whom he may devour" (1 Pet. v. 8.); then changing into a *subtle serpent* (Gen. iii. 8.); then changing again into "an angel of light" (2 Cor. xi. 14.). "Transforming himself" continually, to devour, deceive, and destroy.

This brings us to the conclusion of the Second Book, in which we have seen the Redeemed blessed with all blessings, delivered out of all conflict, saved from all enemies. We have seen their Redeemer, "the Lamb slain from the foundation of the world," "the Conqueror," "the King of Kings and Lord of Lords."

This is the Revelation recorded in the heavens. This is the prophetic testimony inspired in the Book. And this is the heart-cry prompted by both:—

> "Come, Lord, and tarry not,
> Bring the long-looked-for day;
> Oh, why these years of waiting here,
> These ages of delay?
>
> Come, for Thy saints still wait;
> Daily ascends their cry:
> 'The Spirit and the Bride say, Come';
> Dost Thou not hear their cry?
>
> Come, for creation groans,
> Impatient of Thy stay;
> Worn out with these long years of ill,
> These ages of delay.
>
> Come, for Thine Israel pines,
> An exile from Thy fold;
> Oh, call to mind Thy faithful word.
> And bless them as of old.

Come, for Thy foes are strong;
With taunting lips they say,
'Where is the promised advent now,
And where the dreaded day?'

Come, for the good are few;
They lift the voice in vain;
Faith waxes fainter on the earth,
And love is on the wane.

Come, in Thy glorious might;
Come, with Thine iron rod;
Disperse Thy foes before Thy face,
Most mighty Son of God.

Come, and make all things new,
Build up this ruined earth;
Restore our faded paradise,
Creation's second birth.

Come, and begin Thy reign
Of everlasting peace;
Come, take the kingdom to Thyself,
Great King of Righteousness."

Dr. Horatius Bonar.

The Third Book.

THE REDEEMER.

His Second Coming.

IN this Third and Last Book we come to the concluding portion of this Heavenly Revelation. Its subject is Redemption completed, and consummated in triumph. No more sorrow, suffering, or conflict; no more the bruising of the heel of the Redeemer. We have now done with the prophecies of "the sufferings of Christ," and have come to those that relate to "the glory that should follow."

No more reference now to His *first* coming in humiliation. No more coming "forth" to suffer and die, a sacrifice for sins; the reference now is only to His second coming in glory; His coming "unto" this earth is not to suffer for sin (Heb. ix. 28.), but it will be a coming in power to judge the earth in righteousness, and to subdue all enemies under His feet.

Like the other two books, it consists of four chapters.

The *first* chapter is the prophecy of the coming Judge of all the earth.

The *second* sets before us the two-fold nature of the coming Ruler.

The *third* shows us Messiah's redeemed possessions —the Redeemed brought safely home, all conflict over.

The *fourth* describes Messiah's consummated triumph.

CHAPTER I.

The Sign TAURUS (The Bull).

Messiah, the coming Judge of all the Earth.

The picture is that of a Bull rushing forward with mighty energy and fierce wrath, his horns set so as to push his enemies, and pierce them through and destroy them.

It is a prophecy of Christ, the coming Judge, and Ruler, and "Lord of all the earth."

The Egyptian Zodiac of Denderah already, 4,000 years ago, had forgotten the truth to which the prophecy had referred, and called him *Isis*, i.e., *who saves* or *delivers*, and *Apis*, i.e., *the head* or *chief*. The Bull is clearly represented, and in all the zodiacs which have come down to us is always in the *act of pushing*, or *rushing*.

The name of the sign in Chaldee is תּוֹר, *Tor*. Hence, Arabic, *Al Thaur;* Greek, *Tauros*; Latin, *Taurus*, etc. The more common Hebrew name was שׁוֹר, *Shur*, which is from a root which means both *coming* and *ruling*. There are several Hebrew words for bulls and oxen, etc. But the common poetical term for all is ראם, *Reem*, conveying the idea of loftiness, exaltation, power, and pre-eminence. We find the root in other kindred languages (Etruscan, Sanscrit, etc.), and it can be traced in the name of Abram, which means *pre-eminent* or *high father; Ramah*, *high place*, etc.

XXVIII.
TAURUS.
(the Bull)
ARIES
PLEIADES
Asterope
Celino
Merope
Alcyone
Electra
Pleione
Atlas
Maia
HYADES
Al Debaran
El Nath
AURICA
ECLIPTIC
ORION
Bellatrix
Betelgeuse
Menkar
CETUS

The stars in Taurus present a brilliant sight. There are at least 141 stars, besides two important groups of stars, which both form integral parts of the sign.

The brightest star, α (in the bull's eye), has a Chaldee name—*Al Debaran*, and means *the leader* or *governor*. The star β (at the tip of the left horn) has an Arabic name—*El Nath*, meaning *wounded* or *slain*. Another prophetic intimation that this coming Lord should be first slain as a sacrifice.

Then there is the cluster of stars known as the *Pleiades*. This word, which means *the congregation of the judge* or *ruler*, comes to us through the Greek Septuagint as the translation of the Hebrew כִּימָה, *Chima*, which means *the heap* or *accumulation*, and occurs in Job ix. 9; xxxviii. 31, 32, and Amos v. 8.

It consists of a number of stars (in the neck of of Taurus) which appear to be near together. The brightest of them, marked η in all the maps,* has come down to us with an Arabic name—*Al Cyone*, which means *the centre*, and has given the idea to some astronomers that it is the centre of the whole universe. The Syriac name for the Pleiades is *Succoth*, which means *booths*.

Another group of stars (on the face of the Bull) is known as *The Hyades*,† which has the similar meaning of *the congregated*.

---

* The others have names, but they were given by the Greeks from the names of the seven daughters of *Atlas* and *Pleione*. The Hyades were their sisters. Together they tell us that the saints will be secure with this mighty Lord when he comes to rule.

† The Pleiades and Hyades are sometimes spoken of as constellations, but this is a mistake; they are integral parts of Taurus.

Other stars, not identified, are named *Palilicium* (Hebrew), *belonging to the judge; Wasat* (Arabic), *centre* or *foundation; Al Thuraiya* (Arabic), *the abundance; Vergiliæ* (Latin), *the centre* (Arabic, *vertex*) *turned on, rolled round.*

Every thing points to the important truth, and all *turns* on the fact that the Lord is COMING TO RULE! This is the central truth of all prophecy. "The testimony of Jesus is the spirit of prophecy." All hope for Creation, all hope for the world, all hope for Israel, all hope for the Church, turns on this, that "Jesus is coming again," and that when He comes His saints, "the daughters of the King" (like the Pleiades and Hyades), will be with Him.

There is nothing of "the Church" revealed here. The Church will be caught up to meet the Lord in the air, to be for ever with the Lord (1 Thess. iv. 17) *before* He thus *comes unto* the world in judgment. He will *come forth* to receive the members of His Body unto Himself, before He thus comes with them to destroy all His enemies and "judge (or rule) the world in righteousness." When we read this Sign of Taurus, therefore, we are to understand that His Church will be *with* Him, safe from all judgment.

There is very much in the Scripture of the Book, (as there is in the prophecies in the heavens) about the coming of the Lord in judgment; and about this time of His indignation. For Enoch, who doubtless was used in arranging these prophetic *signs*, uttered the prophetic *words*, "Behold the Lord cometh with ten thousands of His saints to execute judgment upon all and to convict all that are ungodly" (Jude 14, 15).

We have said (pages 17, etc.) that at a very early period these signs were appropriated to the Twelve Tribes of Israel, and borne upon their "standards." This may be traced in the Blessing of Jacob (Gen. xlix.), and in the Blessing of Moses (Deut. xxxiii.). Taurus was assigned to Joseph, or rather to his two tribes of Ephraim and Manasseh, like the two powerful horns:

> "The firstling of his bullock (marg. *his firstling bullock*)—
> majesty is his,
> And his horns are the horns of the wild-ox (*Reem*).
> With them he shall PUSH (marg. *gore*) the peoples, all
> of them, even the ends of the earth.
> And they are the ten thousands of Ephraim,
> And they are the thousands of Manasseh."
> (Deut. xxxiii. 17, R.V.)

It is not, however, merely by men alone that this will be done, for David sings:

> "Thou art my King, O GOD. . . .
> Through Thee will we PUSH down our enemies;
> Through Thy Name will we tread them under that
> rise up against us." (Ps. xliv. 5.)

> "I will punish the world for their evil,
> And the wicked for their iniquity;
> I will cause the arrogancy of the proud to cease,
> And will lay low the haughtiness of the terrible. . . .
> Every one that is found shall be THRUST THROUGH."
> (Isa. xiii. 11–15.)

Speaking of that day, the Holy Spirit says by Isaiah:

"For the LORD hath indignation against all the nations,
And fury against all their host:
He hath utterly destroyed them,
He hath delivered them to the slaughter. . . .
The LORD hath a sacrifice in Bozrah,
And a great slaughter in the land of Edom,
And the wild oxen [*Reem*] shall come down with them,
And the bullocks with the bulls;
And their land shall be drunken with blood,
And their dust made fat with fatness.
For it is the day of the LORD's vengeance,
The year of recompense in the controversy of Zion."
(Isa. xxxiv. 2–8, R.V.)

"Behold, the LORD cometh forth out of His place
To punish the inhabitants of the earth for their iniquity:
The earth also shall disclose her blood,
And shall no more cover her slain." (Isa. xxvi. 21.)

This is the united testimony of the two Revelations. It is pictured in the heavens, and it is written in the Book. It is the prophecy of a coming Judge, and of a coming judgment.

It is, however, no mere *Bull* that is coming. It is a man, a glorious man, even "the Son of Man." This is the first development, shown in the first of the three constellations belonging to the sign.

§ 1. ORION (The coming Prince).

Light Breaking Forth in the Redeemer.

This picture is to show that the coming one is no mere animal, but a man: a mighty, triumphant, glorious prince.

He is so pictured in the ancient Denderah Zodiac, where we see a man coming forth pointing to the three bright stars (*Rigel*, *Bellatrix*, and *Betelguez*) as his. His name is given as *Ha-ga-t*, which means *this is he who triumphs*. The hieroglyphic characters below read *Oar*. Orion was anciently spelt *Oarion*, from the Hebrew root, which means *light*. So that Orion means *coming forth as light*. The ancient Akkadian was *Ur-ana*, *the light of heaven*.

Orion is the most brilliant of all the constellations, and when he comes to the meridian he is accompanied by several adjacent constellations of great splendour. There is then above the horizon the most glorious view of the celestial bodies that the starry firmament affords; and this magnificent view is visible to all the habitable world, because the equinoctial line (or solstitial colure) passes nearly through the middle of Orion.

ARATUS thus sings of him:—

"Eastward, beyond the region of the Bull,
Stands great Orion. And who, when night is clear,
Beholds him gleaming bright, shall cast his eyes in vain
To find a Sign more glorious in all heaven."

The constellation is mentioned by name, as being perfectly well known both by name and appearance, in the time of Job; and as being an object of familiar knowledge at that early period of the world's history. See Job ix. 9; xxxviii. 31, and Amos v. 8 (Heb. כְּסִיל, *Chesil*, which means *a strong one*, *a hero*, or *giant*).

It contains 78 stars, two being of the 1st magnitude, four of the 2nd, four of the 3rd, sixteen of the 4th, etc.

A little way below ι (in the sword) is a very remarkable nebulous star. A common telescope will show that it is a beautiful nebula. A powerful telescope reveals it as consisting of collections of nebulous stars, these again being surrounded by faint luminous points, which still more powerful telescopes would resolve into separate stars.

Thus beautifully is set forth the brilliancy and glory of that *Light* which shall break forth when the moment comes for it to be said, "Arise, shine, for thy light is come."

The picture presents us with "the Light of the world." His left foot is significantly placed upon the head of the enemy. He is girded with a glorious girdle, studded with three brilliant stars; and upon this girdle is hung a sharp sword. Its handle proves that this mighty Prince is come forth in a new character. He is again proved to be "the Lamb that was slain," for the hilt of this sword is in the form of the head and body of a lamb. In his right hand he lifts on high his mighty club; while in his left he holds forth the token of his victory—the head and skin of the "roaring lion." We ask in wonder, "Who is this?" * and the names of the stars give us the answer.

The brightest, α (in the right shoulder), is named *Betelgeuz*, which means *the coming* (Mal. iii. 2) *of the branch.*

The next, β (in the left foot), is named *Rigel*, or *Rigol*, which means *the foot that crusheth.* The foot

* See Jer. xxx. 21; and Matt. xxi. 10.

AURIGA
ECLIPTIC
TAURUS
GEMINI
Al Debaran
Betelgeux
Bellatrix
Al Nitak
Al Nitam
Mintaka
CANIS MAJOR
Sirius
ERIDANUS
Rigel
Saiph
Arnebo
LEPUS
XXIX
ORION
(the Glorious One)

is lifted up, and placed immediately over the head of the enemy, as though in the very act of crushing it. Thus, the name of the star bespeaks the act.

The next star, γ (in the left shoulder), is called *Bellatrix*, which means *quickly coming*, or *swiftly destroying*.

The name of the fourth star, δ (one of the three in the belt), carries us back to the old, old story, that this glorious One was once humbled; that His heel was once bruised. Its name is *Al Nitak, the wounded One.** Similarly the star κ (in the right leg) is called *Saiph, bruised*, which is the very word used in Gen. iii. 15, thus connecting Orion with the primeval prophecy. Like Ophiuchus, he has one leg *bruised;* while, with the other, he is *crushing* the enemy under foot.

This is betokened by other stars, not identified, named *Al Rai, who bruises, who breaks* (as in *Cepheus*); and *Thabit* (Hebrew), *treading on.*

Other (Arabic) names relate to His Person: *Al Giauzâ, the branch; Al Gebor, the mighty; Al Mirzam, the ruler; Al Nagjed, the prince; Niphla* (Chaldee), *the mighty; Nux* (Hebrew), *the strong.*

Some names relate to His coming, as *Betelgeuse* and *Bellatrix*, as above; *Heka* (Chaldee), *coming;* and *Meissa* (Hebrew), *coming forth.*

Such is the cumulative testimony of Orion's stars, which, day after day, and night after night, show forth this knowledge. That testimony was after-

* The star ζ (in the belt) is called *Mintaka, dividing*, as a sacrifice. (Lev. viii. 2.)

wards written in the Book. The Prince of Glory, who was once wounded for the sins of His redeemed, is about to rise up and shine forth for their deliverance. Their redemption draweth nigh; for—

> "The LORD shall go forth as a mighty man,
> He shall stir up jealousy like a man of war;
> He shall cry, yea, roar;
> He shall prevail against His enemies.
> I have [*He says*] long time holden my peace;
> I have been still, and refrained myself:
> Now will I cry like a travailing woman;
> I will destroy and devour at once."
>
> (Isa. xlii. 13, 14.)

Then it will be said to His people (and the setting of the prophecy in its beautiful introverted structure shows us the beauty and glory of the truth it reveals): *

a | **Arise,**
  b | Shine; for **thy light** is come,
    c | And **the glory of the LORD** is risen upon thee.
      d | For, behold, the **darkness** shall cover the earth,
      d | And gross **darkness** the people;
    c | But **the LORD** shall arise upon thee, and **His glory** shall be seen upon thee.
  b | And the Gentiles shall come to **thy light,**
a | And kings to the brightness of thy **rising.** (Isa. lx. 1–3.)

This is "the glory of God" which the heavens constantly declare (Ps. xix. 1). They tell of that blessed time when the whole earth shall be filled with His glory (Num. xiv. 21; Isa. xi. 9); when "the glory of the LORD shall be revealed, and all

* Note, that—
In a and *a*, we have the rising of Israel;
In b and *b*, the light that is come upon her;
In c and *c*, the glory of the LORD; and
In d and *d*, the darkness of the world.

flesh shall see it together" (Isa. xl. 5), as all see now the beauty of Orion's glory.

But side by side with the glory which the coming Light of the world shall bring for His people, there is "that wicked," whom the Lord "shall destroy with the brightness of His coming." Hence, as in the concluding chapter (IV.) of the *First* Book (of which this *Third* Book is the expansion) we had in LYRA (*the harp*), as § 1, Praise prepared for the Conqueror; and in ARA (*the burning pyre*), as § 2, Consuming fire prepared for His enemies: so in the *first* chapter of this book, we have in ORION, as § 1, Glory prepared for the Conqueror; and in ERIDANUS, as § 2, the River of wrath prepared for His enemies. This brings us to—

§ 2. ERIDANUS (The River of the Judge).

The River of Wrath Breaking Forth for His Enemies.

It issues forth, in all the pictures, from the down-coming foot of Orion. While others see in it, from the ignorance of fabled story, only "the River Po," or the "River Euphrates," we see in it, from the meaning of its name, and from the significance of its position, *the river of the Judge.*

In the Denderah Zodiac it is a river under the feet of Orion. It is named *Peh-ta-t*, which means *the mouth of the river.*

It is an immense constellation, and our diagram is on a smaller scale than the others (which are all in relative proportion, except where otherwise noted).

According to the Britannic catalogue, it consists of 84 stars; one of the 1st magnitude, one of the 2nd, eight of the 3rd, etc.

The brightest star, α (at the mouth of the river), bears the ancient name of *Achernar*, which is in, as its name means, *the after part of the river.*

The next star, β (at the source of the river), is named *Cursa*, which means *bent down.* The next, γ (at the second bend in the river), is called *Zourac* (Arabic), *flowing*. Other stars, not identified, are *Pheat*, *mouth* (of the river); and *Ozha*, *the going forth.*

Here, then, we have a river flowing forth from before the glorious *Orion.* It runs in a serpentine course towards the lower regions, down, down, out of sight. In vain the sea monster, *Cetus*, strives to stop its flow. It is "the river of the Judge," and speaks of that final judgment in which the wicked will be cast into the lake of fire. It was evidently originally associated with *fire;* for the Greek myths, though gross perversions, still so connect it. According to their fables, something went wrong with the chariot of the sun, and a universal conflagration was threatened. In the trouble, *Phaeton* (probably a reference to the star *Pheat*) was killed and hurled into this river, in which he was consumed with its fire. The whole earth suffered from such a burning heat that great disasters ensued. We see from this myth two great facts preserved in the perverted tradition, *viz.*, *judgment* and *fire.*

Aratus also preserves the connection,—

"For yonder, trod by heavenly feet,
Wind the scorched waters of Eridanus' tear-swollen flood,
Welling beneath Orion's uplifted foot."

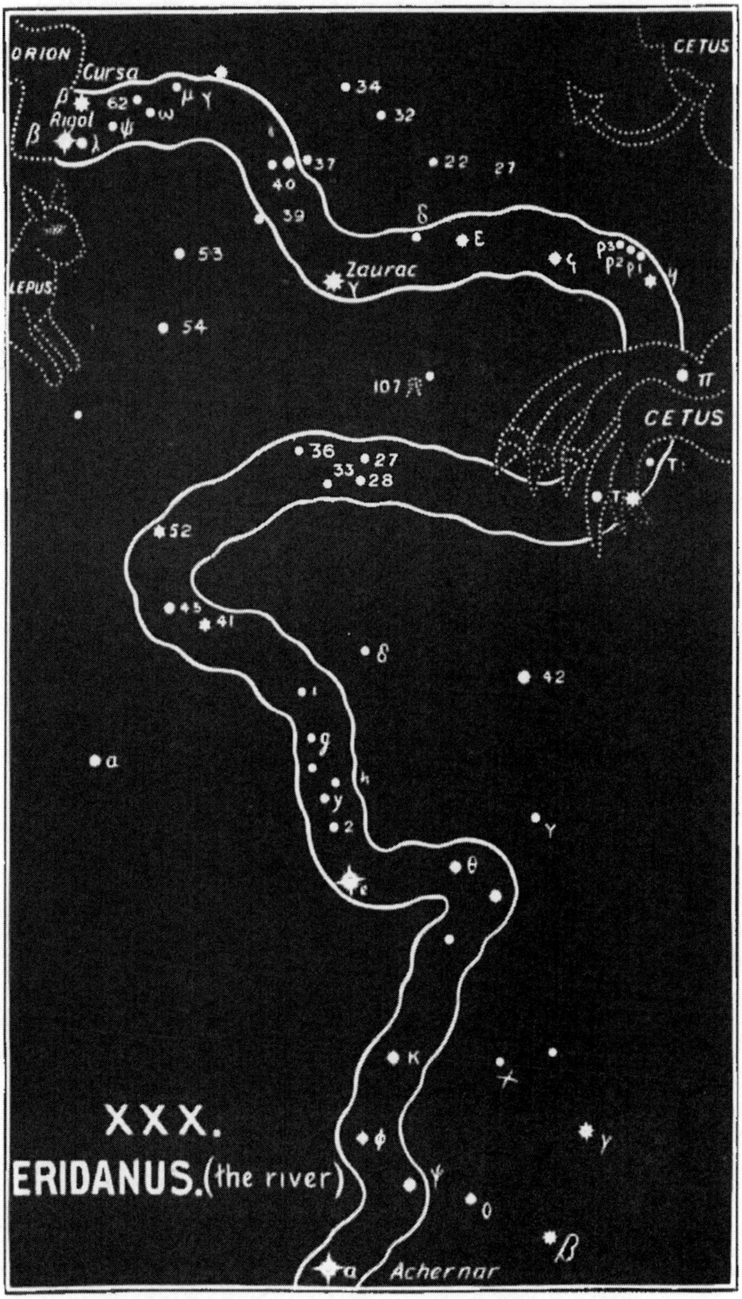

XXX.
ERIDANUS. (the river)

Is not this the testimony afterwards written in *the Book?* Daniel sees this very river in his vision of that coming day, when the true Orion shall come forth in His glory. He says, "I beheld till the thrones were placed, and one that was ancient of days did sit: .... His throne was fiery flames, and the wheels thereof burning fire. A FIERY STREAM ISSUED AND CAME FORTH FROM BEFORE HIM." This is *the River of the Judge;* for he goes on to say, "the judgment was set, and the books were opened" (Dan. vii. 9–11, R.V.).

We have the same in Ps. xcvii. 3–5 (R.V.), which describes the scene when the Lord shall reign:

"A FIRE GOETH BEFORE HIM,
And burneth up His adversaries round about.
His lightnings lightened the world:
The earth saw and trembled,
The hills melted like wax at the presence of the LORD,
At the presence of the Lord of the whole earth."

So again in Ps. l. 3, we read:

"Our God shall come, and shall not keep silence,
A FIRE SHALL DEVOUR BEFORE HIM,
And it shall be very tempestuous round about Him."

By Habakkuk the coming of the Lord is described; and it is written:

"His brightness was as the light, ...
Before Him went the pestilence,
And burning coals went forth at His feet."
(Hab. iii. 5.)

What is this but *Orion* and *Eridanus!*

Again, it is written in Isaiah xxx. 27–33 (R.V.):

"Behold, the name of the LORD cometh from far,
Burning with His anger, and in thick rising smoke:

His lips are full of indignation,
And His tongue is as a DEVOURING FIRE:
And His breath is as AN OVERFLOWING STREAM [*of fire*]. . . .
For a Topheth is prepared of old;
Yea, for the king [*Moloch*] it is made ready;
He hath made it deep and large;
The pile thereof is FIRE and much wood;
The breath of the LORD, LIKE A STREAM OF BRIMSTONE,
doth kindle it."

So, again, we read in Nahum i. 5, 6:

"The mountains quake at Him,
And the hills melt;
And the earth is burned up at His presence,
Yea, the world and all that dwell therein.
Who can stand before His indignation?
And who can abide in the fierceness of His anger?
His fury is POURED OUT LIKE FIRE."

In Isa. lxvi. 15, 16, we read:

"For, behold, the LORD will come with fire,
And with His chariots like a whirlwind,
To render His anger with fury,
And His rebuke with FLAMES OF FIRE,
For BY FIRE, and by His sword, will the LORD plead with
all flesh."

With this agree the New Testament scriptures, which speak of "the Day of the Lord," "when the Lord Jesus shall be revealed from heaven with His mighty angels, IN FLAMING FIRE taking vengeance on them that know not God, and that obey not the Gospel of our Lord Jesus Christ" (2 Thess. i. 7, 8).

This is the true Eridanus. It is no mere "picture." It is a dread reality! It is written in stars of fire, and words of truth, that men may heed the solemn warning and "flee from the wrath to come"!

But we ask, "Who may abide the day of His coming? and who shall stand when He appeareth" (Mal. iii. 2)? "Who can stand before His indignation," when "His fury is poured out like fire" (Nah. i. 6)?

The answer is given in the next picture!

§ 3. AURIGA (The Shepherd).

Safety for the Redeemed in the Day of Wrath.

Here is presented to us the answer to the question, "Who may abide the day of His coming?"

> "Behold, the Lord God (Adonai Jehovah) will come
> as a mighty one,
> And His arm shall rule for Him:
> Behold, His reward is with Him,
> And His recompense before Him.
> He shall feed His flock like a shepherd,
> He shall gather the lambs in His arm,
> And carry them in His bosom,
> And shall gently lead those that give suck."
> (Isa. xl. 10, 11, R.V.)

This is exactly what is presented before us in this last section of the chapter, which tells of the coming judgment. We have had the picture of a mighty *Bull* rushing forth; then the fiery river of *the Judge;* and now we see *a Great Shepherd.* He is seated upon "the milky way," holding up on his left shoulder a she goat. She clings to his neck, and is looking down affrighted at the terrible onrushing Bull. In his left hand he supports two little kids, apparently just born, and bleating, and trembling with fear.

ARATUS says,—

> "She is both large and bright, but they—the kids—
> Shine somewhat feebly on *Auriga's* wrist."

Is not this the Great Shepherd gathering the lambs in His arm? and carrying them in His bosom? Is He not saying:

> "I will save My flock,
> And they shall no more be a prey."
> (Ezek. xxxiv. 22.)

> "And David my servant shall be king over them,
> And they shall have one shepherd." (*ib.* xxxvii. 24.)

> "And they shall fear no more,
> Nor be dismayed,
> Neither shall they be lacking, saith the LORD."
> (Jer. xxiii. 4.)

AURIGA is from a Hebrew root which means *a shepherd.* It is a beautiful constellation of 66 stars; one of the 1st magnitude, two of the 2nd, nine of the 4th, etc.

The brightest star, α (in the body of the goat), points her out as the prominent feature of the constellation, for its name *Alioth* (Hebrew) means *a she goat.* It is known by the modern Latin name *Capella*, which has the same meaning.

The next star, β (in the shepherd's right arm), is called *Menkilinon*, and means the *band*, or *chain of the goats*, and points out the truth that they are never more to be lost again, but to be bound, with the bands of love, to the Shepherd for evermore.

XXXI. AURIGA (the Shepherd)

The name of another star is *Maaz*, which means *a flock of goats*.

Can there be any mistake as to who this Shepherd is? for the bright star in his right foot is called *El Nath* * (like another in ARIES), which means *wounded* or *slain*. This is He, then, who was once bruised or wounded in the heel. He is "the GOOD Shepherd," who gave His life for the sheep (John x. 11), but He was "the GREAT Shepherd" brought again from the dead (Heb. xiii. 20); and is now the CHIEF Shepherd (1 Pet. v. 4) seen in the day of His coming glory. Another star emphasises this truth, for it is named *Aiyuk*, which also means *wounded* in the foot.†

The star marking the kids is called *Gedi* (Hebrew), *kids*.

In Latin, the word *Auriga* means a *coachman* or *charioteer*, the band in his right hand being taken as his *reins*. But the incongruity of a *charioteer* carrying a she-goat, and nursing two little kids, never struck them; nor did the fact that he has no chariot and no horses! When man blunders in the things of God, he does it thoroughly!

In the Zodiac of Denderah the same truth was revealed more than 4,000 years ago; but the Man, instead of carrying the sheep, is carrying a sceptre, and is called *Trun*, which means *sceptre* or *power*. But this is a strange sceptre, for at the top it has the head of a goat, and at the bottom, below

* It is also reckoned in the horn of Taurus.

† The same as in 2 Sam. iv. 1.

the hand that holds it, it ends in a cross! With the Egyptians the cross was a sign of *life*. They knew nothing of "the death of the cross." Here, then, we see *life* and *salvation* for the sheep of His flock when He comes to reign and rule in judgment. The truth is precisely the same, though the presentation of it is somewhat varied.

The connected teaching of the two constellations, *Eridanus* and *Auriga*, is solemnly set forth in Mal. iv. 1–3 (R.V.):

> "Behold, the day cometh,
> It burneth as a furnace;
> And all the proud, and all that work wickedness, shall be stubble:
> And the day that cometh shall burn them up, saith the LORD of hosts,
> That it shall leave them neither root nor branch.
> BUT UNTO YOU that fear My name shall the Sun of Righteousness arise with healing in His wings;
> And ye shall go forth and gambol as calves of the stall.
> And ye shall tread down the wicked;
> For they shall be ashes under the soles of your feet
> In the day that I do make (marg. *do this*), saith the LORD of hosts."

In Psalm xxxvii. this day is repeatedly referred to, the day when "the wicked shall be cut off"; and it concludes by summarizing the same great truth (*vv.* 38–40, R.V.):

> "As for transgressors, they shall be destroyed together;
> The latter end of the wicked shall be cut off,
> But the salvation of the righteous is of the LORD:
> He is their stronghold in the time of trouble,
> And the LORD helpeth them, and rescueth them;
> He rescueth them from the wicked and saveth them,
> Because they have taken refuge in Him."

Oh, that all who read these pages may heed the solemn warning, and flee for refuge to Him who now, in this day of grace, is crying, " Look unto me, and be ye saved, O all ye ends of the earth " (Isa. xlv. 22).

CHAPTER II.

THE SIGN GEMINI (The Twins).

Messiah's Reign as Prince of Peace.

ALL the pictures of this sign are confused. The Greeks claimed to have invented them, and they called them Apollo and Hercules. They are so given in our illustration. The Latins called them Castor and Pollux; and the name of a vessel in which Paul sailed is so called in Acts xxviii. 11, Διοσκοῦροι.

The name in the ancient Denderah Zodiac is *Clusus*, or *Claustrum Hor:*, which means *the place of Him who cometh.* It is represented by two human figures walking, or coming. The second appears to be a woman. The other appears to be a man. It is a tailed figure, the tail signifying *He cometh.*

The old Coptic name was *Pi-Mahi*, *the united*, as in brotherhood. Not necessarily united by being born at the same time, but *united* in one fellowship or brotherhood. The Hebrew name is *Thaumim*, which means *united.* The root is used in Exod. xxvi. 24:

"They (the two boards) shall be coupled together beneath." In the margin we read, "Heb. *twinned*" (R.V. double). The Arabic *Al Tauman* means the same.

We need not trouble ourselves with the Grecian myths, even though we can see through them the original and ancient truth. The two were both heroes of peculiar and extraordinary birth—sons of Jupiter. They were supposed to appear at the head of armies; and as they had cleared the seas of pirates, they were looked upon as the patron saints of navigation. (Hence the name of the ship in Acts xxviii. 11.) They were held in high esteem both by Greeks and Romans; and the common practice of taking oaths and of swearing by their names has descended even to our own day in the still surviving vulgar habit of swearing "By Gemini!"

The more ancient star-names help us to see through all these and many other myths, and to discern Him of whom they testify; even Him in His twofold nature—God and Man—and His twofold work of suffering and glory, and His twofold coming in humiliation and in triumph.

There are 85 stars in the sign: two of the 2nd magnitude, four of the 3rd, six of the 4th, etc.

The name of α (in the head of the one at our right hand) is called *Apollo*, which means *ruler*, or *judge;* while β (in the head of the other) is called *Hercules, who cometh to labour*, or *suffer*. Another star, γ (in his left foot), is called *Al Henah*, which

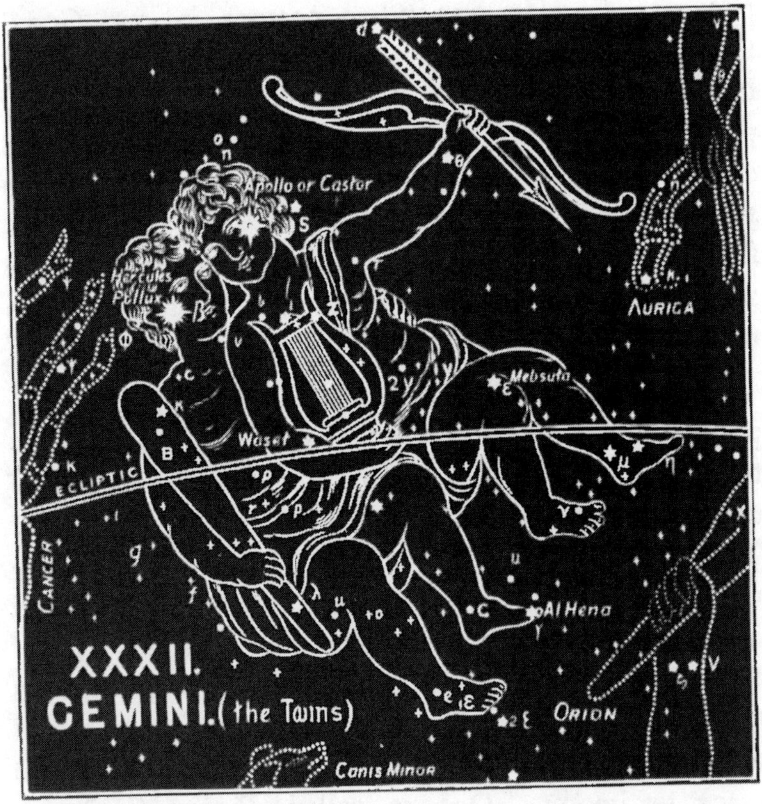

XXXII.
GEMINI. (the Twins)

means *hurt, wounded*, or *afflicted*. Can we have a doubt as to what is the meaning of this double presentation? In *Ophiuchus* we have the two in one person: the crushed enemy, and the wounded heel. But here the two great primeval truths are presented in two persons; for He was "two persons in one God," "God and man in one Christ." As man, suffering for our redemption; as God, glorified for our complete salvation and final triumph. A star, ε (in the centre of his body), is called *Waset*, which means *set*, and tells of Him who "set His face like a flint" to accomplish this mighty Herculean work; and, when the time was come, "steadfastly set His face to go" to complete it.

He bears in his right hand (in some pictures) a palm branch. In the one from which our illustration is taken, it is a club; but both the club of this one and the bow of the other are *in repose!* These united ones are neither in action nor are they preparing for action, but they are *at rest* and *in peace* after victory won. The star ε (in the knee of the other, "Apollo") is called *Mebsuta*, which means *treading under feet*. The names of other stars, which are not identified, have come down to us with the same testimony. One is called *Propus* (Hebrew), *the branch, spreading;* another is called *Al Giauzâ* (Arabic), *the palm branch;* another is named *Al Dirâ* (Arabic), *the seed*, or *branch*.

The day has here come to fulfil the prophecies concerning Him who is "the Branch," "the Branch of Jehovah," "the man whose name is the Branch."

"In that day shall the Branch of Jehovah be beautiful
and glorious;
And the fruit of the earth shall be excellent and comely
For them that are escaped of Israel." (Isa. iv. 2.)

"Behold, a king shall reign in righteousness,
And princes shall rule in judgment;
And a man shall be as an hiding place from the wind."
(Isa. xxxii. 1, 2.)

"Behold, the days come, saith the LORD,
That I will raise unto David a righteous Branch,
And He shall reign as King and deal wisely,
And shall execute judgment and justice in the land.
In His days Judah shall be saved,
And Israel shall dwell safely:
And this is His name whereby He shall be called,
The LORD is our Righteousness." (Jer. xxiii. 5, 6, R.V.)

"Behold, the days come, saith the LORD,
That I will perform that good word which I have
spoken
Concerning the house of Israel and concerning the
house of Judah.
In those days, and at that time,
Will I cause a Branch of Righteousness to grow up
unto David;
And He shall execute judgment and righteousness in
the land." (Jer. xxxiii. 14, 15, R.V.)

This is what we see in this sign—Messiah's peaceful reign. All is rest and repose. We see "His days," in which "the righteous shall flourish; and abundance of peace, so long as the moon endureth" (Ps. lxxii.).

But, for this blessed time to come, there must be no enemy! All enemies must be subdued.

This brings us to the first section of this book.

XXXIII. LEPUS. (the Hare) or ENEMY

§ 1. LEPUS (the Hare), THE ENEMY.

The Enemy Trodden under Foot.

The names of the three constellations of this Sign, as well as the pictures, are all more or less modern, as is manifest from the names being in *Latin*, and having no relation to the ancient names of their stars. To learn their real meaning, therefore, we must have recourse to the ancient Zodiacs. In the Persian planisphere the first constellation was pictured by *a serpent*. In the Denderah (Egyptian) Zodiac it is an unclean bird standing on the serpent, which is under the feet of Orion. Its name there is given as *Bashti-beki*. *Bashti* means *confounded*, and *Beki* means *failing*.

ARATUS says,—

> "Below Orion's feet, the Hare
> Is chased eternally."

It is a small constellation of 19 stars (all small), three of which are of the 3rd magnitude, seven of the 4th, etc.

The brightest, α (in the body), has a Hebrew name, *Arnebo*, which means *the enemy of Him that cometh*. The Arabic, *Arnebeth*, means the same. Other stars, not identified, are *Nibal*, *the mad; Rakis*, *the bound* (Arabic, *with a chain*); *Sugia*, *the deceiver*.

There can be no mistaking the voice of this united testimony. For this enemy is under the down-coming foot of Orion, and it tells of the blessed fact that when the true Orion, "the Sun of

Righteousness, shall arise," and "the true light" shall shine over all the earth, He "shall tread down the wicked" (Mal. iv.), and every enemy will be subdued under His feet. "It is He that shall tread down our enemies" (Ps. lx. 12), as He has said:

> "I will tread them in Mine anger,
> And trample them in My fury . . .
> For the day of vengeance is in Mine heart,
> And the year of My redeemed is come."
> (Isa. lxiii. 3, 4.)

§ 2. CANIS MAJOR (The Dog), or SIRIUS (The Prince).

The coming Glorious Prince of Princes (Sirius).

This second constellation carries on the teaching, and tells of the glorious Prince who will thus subdue and reign.

In the Denderah Zodiac he is called *Apes*, which means *the head.* He is pictured as a hawk (*Naz*, נץ, *caused to come forth*, *coming swiftly down*). The hawk is the natural enemy of the serpent, and here it has on its head a pestle and mortar, indicating the fact that he shall crush the head of the enemy.

In the Persian planisphere it is pictured as *a wolf*, and is called *Zeeb*, which in Hebrew (זְאֵב) has the same meaning. Plutarch translates it Προόπτης, *Leader*. In Arabic it means *coming quickly*.

Its ancient name and meaning must be obtained from the names of its stars which have come down

XXXIV. CANIS MAJOR (the Dog)
CANIS MINOR. (the second Dog)

to us. There are 64 altogether. Two are of the 1st magnitude, two of the 2nd, four of the 3rd, four of the 4th, etc. Of these α (in the head) is the brightest in the whole heavens! It is called *Sirius, the Prince* (שַׂר, *Sar*), as in Isa. ix. 6.

Sirius * was, by the ancients, always associated with great heat. And the hottest part of the year we still call "the dog days," though, through the variation as observed in different latitudes, and the precession of the equinoxes, its rising has long ceased to have any relation to those days. Virgil says that Sirius

"With pestilential heat infects the sky."

Homer spoke of it as a star

"Whose burning breath
Taints the red air with fevers, plagues, and death."

It is not, however, of its heat that its name speaks, but of the fact that it is the brightest of all the stars, as He of whom it witnesses is the "Prince of princes," "the Prince of the Kings of the earth."

Though this "Dog-Star" came to have an ill-omened association, it was not so in more ancient times. In the ancient Akkadian it is called *Kašišta*, which means *the Leader* and Prince of the heavenly host. While (as Mr. Robert Brown, Junr., points out) "the Sacred Books of Persia contain many praises of the star *Tistrya* or *Tistar* (*Sirius*), 'the chieftain of the East.'"†

---

* Our English "Sir" is derived from this word.

† *Euphratean Stellar Researches.*

The next star, β (in the left fore foot), speaks the same truth. It is named *Mirzam*, and means *the prince* or *ruler*. The star δ (in the body) is called *Wesen, the bright, the shining*. The star ε (in the right hind leg) is called *Adhara, the glorious.*

Other stars, not identified, bear their witness to the same fact. Their names are—*Aschere* (Hebrew), *who shall come; Al Shira Al Jemeniya* (Arabic), *the Prince* or *chief of the right hand! Seir* (Egyptian), *the Prince; Abur* (Hebrew), *the mighty; Al Habor* (Arabic), *the mighty; Muliphen* (Arabic), *the leader, the chief.*

Here there is no conflicting voice; no discord in the harmonious testimony to Him whose name is called "Wonderful, Counsellor, the Mighty God . . . the Prince of Peace" (Isa. ix. 6).

The names of the stars have no meaning whatever as applied to an Egyptian Hawk, or a Greek Dog. But they are full of significance when we apply them to Him of whom Jehovah says:

> "Behold, I have given Him for a witness to the people,
> A LEADER and commander to the people."
>
> (Isa. lv. 4.)

This is "the Prince of princes" (Dan. viii. 23, 25) against whom, "when transgressors are come to the full, a king of fierce countenance . . . shall stand up," "but he shall be broken without hand," for he shall be destroyed "with the brightness of His coming" (2 Thess. ii. 8). This is He who shall come forth "King of kings and Lord of Lords" (Rev. xix. 16).

But Sirius has a companion, and this brings us to—

§ 3. CANIS MINOR (The Second Dog).

The Exalted Redeemer (Procyon).

The same facts are to be remembered concerning the Greek picture, and Latin name of this constellation.

The Egyptian name in the Denderah Zodiac is *Sebak*, which means *conquering, victorious.* It is represented as a human figure with a hawk's head and the appendage of a tail.

This small constellation has only 14 stars according to the Britannic catalogue. One of the 1st magnitude, one of the 2nd, one of the 4th, etc.

The brightest star, α (in the body), is named *Procyon*, which means REDEEMER, and it tells us that this glorious Prince is none other than the one who was slain. Just as this chapter begins with *two* persons in one in the Sign (Gemini), one *victorious*, the other *wounded;* so it ends with a representation of two princes, one of whom is seen triumphant and the other as the Redeemer. This is confirmed by the next star, β (in the neck), which is named *Al Gomeisa* (Arabic), *the burthened, loaded, bearing for others.* The names of the other stars, not identified, still further confirm the great truth; viz., *Al Shira*, or *Al Shemeliya* (Arabic), *the prince or chief of the left hand*, answering to the star in *Sirius.* One *right*, the other *left*, as the two united youths are placed. *Al*

Mirzam, the prince or *ruler;* and *Al Gomeyra, who completes* or *perfects.*

This does, indeed, complete and perfect the presentation of this chapter: Messiah's reign as Prince of Peace; the enemy trodden under foot by the glorious "Prince of princes," who is none other than the glorified Redeemer.

This is also what is written in the Book:

"Shall the prey be taken from the mighty,
Or the lawful captives* be delivered?
But thus saith the LORD,
Even the captives of the mighty shall be taken away,
And the prey of the terrible shall be delivered:
For I will contend with him that contendeth with thee,
And I will save thy children.
And I will feed them that oppress thee with their own flesh;
And they shall be drunken with their own blood as with sweet wine;
And all flesh shall know that I the LORD am thy Saviour,
And thy REDEEMER—the Mighty One of Jacob."
(Isa. xlix. 24–26, R.V.)

"When the enemy shall come in like a flood,
The Spirit of the LORD shall lift up a standard against him,
And the REDEEMER shall come to Zion."
(Isa. lix. 19, 20.)

"And He shall divide the spoil with the strong,
Because He hath poured out His soul unto death."
(Isa. liii. 12.)

---

* Marg., "*the captives of the just,*" or, as read by the Vulg. and Syr., "*the captives of the terrible.*"

CHAPTER III.

THE SIGN CANCER (The Crab).

Messiah's Redeemed Possessions held fast.

WITH regard to the sign of CANCER, one thing is certain, that we have not got the original picture, or anything like it.

It does not agree with the names either of its three constellations which have come down to us, or of its stars.

In the ancient Denderah Zodiac it is represented as a *Scarabæus*, or sacred beetle.* In the Zodiac of Esneh and in a Hindu Zodiac (400 B.C.) it is the same.

According to the Greeks, Jupiter placed this Crab amongst the signs of the Zodiac.

In Sir William Jones's Oriental Zodiac we meet with a crab, and an Egyptian Zodiac found at Rome bears also the crab in this sign.

The more ancient Egyptians placed *Hermanubis*, or *Hermes*, with the head of an ibis or hawk, as the symbol of the sign now allotted to Cancer.

The Denderah name is *Klaria*, or *the cattle-folds*, and in this name we have the key to the meaning of the sign, and to the subject of this chapter.

---

* The Scarabæus, passing its early existence as a worm of the earth, and thence issuing as a winged denizen of heaven, was held sacred by the Egyptians as an emblem of the resurrection of the body.

The Arabic name is *Al Sartan*, which means *who holds* or *binds*, and may be from the Hebrew אָסַר, *to bind together* (Gen. xlix. 11). There is no ancient Hebrew word known for the crab. It was classed with many other unclean creatures, and would be included in the general term "vermin."

The Syriac, *Sartano*, means the same. The Greek name is *Karkinos*, which means *holding* or *encircling*, as does the Latin, *Cancer*, and hence is applied to the crab. In the word *Khan*, we have the traveller's rest or *inn;* while *Ker* or *Cer* is the Arabic for *encircling*. The ancient Akkadian name of the month is *Su-kul-na*, *the seizer* or *possessor of seed*.

The sign contains 83 stars, one of which is of the 3rd magnitude, and seven are of the 4th magnitude, and the remainder of inferior magnitudes.

In the centre of the Sign there is a remarkably bright cluster of stars, so bright that they can be sometimes seen with the naked eye. It looks like a comet, and is made up of a great multitude of stars. Modern astronomers have called it the *Bee-hive.* But its ancient name has come down to us as *Praesepe*, which means *a multitude, offspring*.

The brightest star, ζ (in the head), is called *Tegmine*, *holding*. The star α (or α^1 and α^2), in the lower large claw, is called *Acubene*, which, in Hebrew and Arabic, means *the sheltering* or *hiding-place.* Another is named *Ma'alaph* (Arabic), *assembled thousands; Al Himarein* (Arabic), *the kids* or *lambs.*

North and south of the nebula *Praesepe* are two stars, which Orientalists speak of by a name evidently

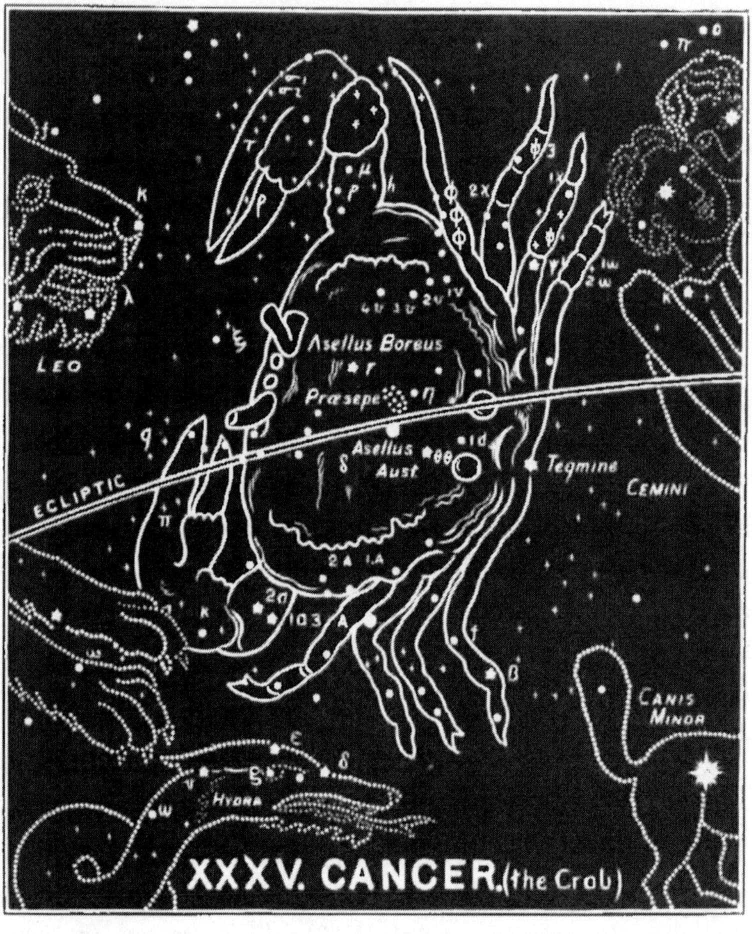

XXXV. CANCER. (the Crab)

of some antiquity. *Asellus* means *an Ass*, and one was called *Asellus Boreas, the northern Ass;* while the other, *Asellus Australis*, is *the southern Ass.*

The sign was afterwards known by the symbol ♋, which stands for these two asses.*

This connects it with the Tribe of Issachar, who is said to have borne upon the Tribal standard the sign of *two asses*.

This is doubtless the reference in Jacob's blessing (Gen. xlix. 11, R.V.) :—

> "Issachar is a strong ass,
> Couching down between the sheepfolds;
> And he saw a resting-place that it was good;
> And the land that it was pleasant;
> And he bowed his shoulder to bear,
> And became a servant under task work."

Have we not here the gathering up of the teaching of this sign—

Messiah's redeemed possessions held fast.

Here we come to the completion of His work. In CANCER we see it with reference to His *redeemed*, and in the next (the last) Sign, LEO, with reference to His *enemies*.

The three constellations develope the truth. What is now called *Ursa Minor* is *the Lesser Flock; Ursa Major* gives us *The Sheepfold and the Sheep;* while *Argo, The Ship*, shows the travellers and the pilgrims brought safely home—all conflict over.

To accomplish this, we see the true Issachar bowing his shoulder to bear. HE could say, "My

* The *Ass* was the emblem of *Typhon*, the king *who smites* or *is smitten*.

soul is bowed down" (Ps. lvii. 6). HE became a servant, and humbled Himself to death. HE undertook the mighty task of saving His people from their sins. "Their Redeemer is strong" (Jer. l. 34); for help was laid on "One that was mighty" (Ps. lxxxix. 19). And His redeemed shall come to a resting-place that is good, and to a land that is pleasant. No earthly Khan on earth affords them a home. They look for a heavenly home, and in the many mansions of the Father's house they shall find eternal rest.

Here we see that sheltering home to which the names of these stars point; where the assembled thousands (*Ma'alaph*) shall be received into the true *Klaria*, even the "everlasting habitations."

These are now to be shown to us.

§ 1. URSA MINOR (The Little Bear).

The Lesser Sheepfold.

Here we come to another grievous mistake, or ignorant perversion of primitive truth, as shown in the ancient names of these two constellations.

It is sufficient to point to the fact that no Bear is found in any Chaldean, Egyptian, Persian, or Indian Zodiacs, and that no bear was ever seen with such a tail! No one who had ever seen a bear would have called attention to a tail, such as no bear ever had, by placing in its very tip the most important, wondrous, and mysterious Polar Star, the central star of the heavens, round which all others revolve. The patriarchal astronomers, we may be sure, committed no such folly as this.

The primitive truth that there were *two*, or a pair of constellations is preserved; and that of these two, one is larger, and the other smaller. But What were they? We have the clue to the answer in the name of the brightest star of the larger constellation, which is called *Dubheh*. Now *Dubheh* means *a herd of animals*. In Arabic, *Dubah* means *cattle*. In Hebrew, דֹּבֶר, *Dōhver*, is *a fold;* and hence in Chaldee it meant *wealth*. The Hebrew דֹּבֶא, *Dōhveh*, means *rest* or *security;* and certainly there is not much of either to be found or enjoyed with bears! The word occurs in Deut. xxxiii. 25: "As thy days so shall thy strength be." The Revised Version gives in the margin, "So shall *thy rest* or *security* be." This accords with what we have already seen under "CANCER": "Couching down between the sheepfolds,* he saw a resting-place that it was good."

Here are the two Sheepfolds, then; the Greater fold, and Lesser; and here is the *rest* and *security* which the flocks will find therein.

But in Hebrew there is a word very similar in sound, though not in spelling—דֹּב or דּוֹב, *dōhv*, which means *a bear!* So we find in Arabic *dub;* Persian, *deeb* and *dob*. We can see, therefore, how the Hebrew *Dōhver*, *a fold*, and *Dōhv*, *a bear*, were confused; and how the Arabic *Dubah*, *cattle*, might easily have been mistaken by the Greeks, and understood as a bear.

The constellation, which we must therefore call THE LESSER SHEEPFOLD, contains 24 stars, *viz.*, one of the 2nd magnitude, two of the 3rd, four of the 4th, etc.

* The word is so rendered in Judges v. 16, in A.V.

The brightest star, α (at the point of the tail), is the most important in the whole heavens. It is named *Al Ruccaba*, which means *the turned* or *ridden on*, and is to-day the Polar or central star, which does not revolve in a circle as does every other star, but remains, apparently, fixed in its position. But though the star does not revolve like the others, the central point in the heavens is very slowly but steadily moving. When these constellations were formed the Dragon possessed this important point, and the star α, in *Draco*, marked this central point. But, by its gradual recession, that point is sufficiently near this star *Ruccaba*, in *the Lesser Sheepfold*, for it to be what is called "the Polar Star." But, how could this have been known five or six thousand years ago? How could it have been known when it received its name, which means *the turned* or *ridden on?* That it was known is clear: so likewise was it made known in the written Word that the original blessing included not merely the multiplication of the seed of faithful Abraham, but it was then added, "And thy seed shall possess the gate of his enemies" (Gen. xxii. 17).

This star was called by the Greeks Κυνόσουρα, the "*Cynosure*." ARATUS seems to apply this term to the whole of the seven stars of the *Lesser Bear*. Mr. Robert Brown, Junr., shows* that Κυνόσουρα, once supposed to be Hellenic, is non-Hellenic, and possibly Euphratean in origin, from a word which he transliterates *An-nas-sur-ra*, and renders it, "as it literally means, *high in* rising, *i.e.*, in heavenly position." Is

* *Euphratean Stellar Researches*, pp. 8, 9.

XXXVI. URSA. MINOR.
Polar Star
Al Ruccaba
or Arcos
Kochab
CEPHEUS
DRACO

not this the primitive truth of the Revelation? Will not this Lesser Fold be high, yea, the highest in heavenly position?

The Polar Star has been removed from the Dragon, and is now in *the Lesser Fold;* and when the Dragon shall be cast down from the heavens, the heavenly seed will be safely folded there. But this is *the Lesser Sheepfold.* These are they who all through the ages have been "partakers of the heavenly calling," who desired a better country, that is, a *heavenly;* wherefore God "hath prepared for them a city," the city for which Abraham himself "looked." This was no earthly city, but a city "whose builder and maker is God" (Heb. xi. 10–16). These have always been a smaller company, a "little flock," but the kingdom shall be theirs, even the kingdom of God, for which they now look and wait. They have not yet "received the promises; but, having seen them afar off" by faith, they "were persuaded of them, and embraced them, and confessed that they were strangers and pilgrims on the earth" (Heb. xi. 13). Their Messiah has accomplished "the redemption of the purchased possession," and in due time the redeemed will inherit it, "unto the praise of His glory" (Eph. i. 13).

The bright star β is named *Kochab*, which means *waiting Him who cometh.* Other stars, not identified, are named *Al Pherkadain* (Arabic), which means *the calves*, or *the young* (as in Deut. xxii. 6), *the redeemed assembly.* Another, *Al Gedi*, means *the kid.* Another is *Al Kaid*, *the assembled;* while *Arcas*, or *Arctos* (from which we derive the term *Arctic* regions), means, according to one interpreter, *a travelling company;* or, according to another, *the stronghold of the saved.*

But there is not only the heavenly seed, which is compared "to the stars of heaven," but there is the seed that is compared to "the sand of the sea" —the larger flock or company, who will enjoy the earthly blessing.

This brings us to—

§ 2. URSA MAJOR (The Great Bear).

The Fold and the Flock.

Of these it is written:—

> "But in Mount Zion there shall be those that escape,
> And it shall be holy:
> And the house of Jacob shall possess their possessions."
> (Obad. 17–19, R.V.)

It is a large and important constellation, containing 87 stars, of which one is of the 1st magnitude, four of the 2nd, three of the 3rd, ten of the 4th, etc. It always presents a splendid appearance, and is perhaps, therefore, the best known of all the constellations.

In the Book of Job (ix. 9, and xxxviii. 31, 32) it is mentioned under the name of *Ash.* "Canst thou guide *Ash* and her offspring?" which is rendered in the A.V., "Arcturus and his sons," and in the R.V., "The Bear with her train" (marg., "*sons*").

XXXVII. URSA MAJOR.

The Arabs still call it *Al Naish*, or *Annaish*, *the assembled together*, as sheep in a fold. The ancient Jewish commentators interpreted *Ash* as the seven stars of this constellation. They are called by others *Septentriones*, which thus became the Latin word for *North*.

The brightest star, α (in the back), is named *Dubhe*, which, as we have seen, means *a herd of animals*, or *a flock*, and gives its name to the whole constellation.

The star β (below it) is named *Merach* (Hebrew), *the flock* (Arabic, *purchased*).

The star γ (on the left of β) is called *Phaeda*, or *Phacda*, meaning *visited*, *guarded*, or *numbered*, as a flock; for His sheep, like the stars, are both *numbered* and *named*. (See Psalm cxlvii. 4.)

The star ε is called *Alioth*, a name we have had in *Auriga*, meaning *a she goat*.

The star ζ (in the middle of the tail) is called *Mizar*, *separate* or *small*, and close to it *Al Cor*, *the Lamb* (known as "g").

The star η (at the end of the so-called tail) is named *Benet Naish* (Arabic), *the daughters of the assembly*. It is also called *Al Kaid*, *the assembled*.

The star ι (in its right foot) is called *Talitha*.

The names of other stars, not identified, all give the same testimony: *El Alcola* (Arabic), *the sheepfold* (as in Ps. xcv. 7; and c. 3); *Cab'd al Asad*, *multitude*, *many assembled; Annaish*, *the assembled; Megrez*, *separated*, as the flock in the fold; *El Kaphrah*, *protected*, *covered* (Heb. *redeemed* and *ransomed*); *Dubheh Lachar* (Arabic), *the latter herd* or *flock; Helike* (so called by HOMER in the *Iliad*), *company of travellers; Amaza* (Greek), *coming and going; Calisto, the sheepfold set* or *appointed.*

There is not one discordant voice in the rich abundance of this testimony. We have nothing to do here with the Grecian myths about bears or wild boars. We see only the innumerable seed *gathered* by Him who *scattered* (Jer. xxxi. 10).

Many are the Scriptures we might quote which speak of this gathering and assembling of the long-scattered flock. It is written as plainly in the Book, as it is in the heavens. The prophecies of this gathering are as conspicuous in the Word of God as the "*Seven Stars*" are in the sky. It is difficult even to make a selection from the wealth of such promises; but few are more beautiful than that in Ezek. xxxiv. 12–16:

> "As a shepherd seeketh out his flock
> In the day that he is among his sheep that are scattered;
> So will I seek out my sheep,
> And will deliver them out of all places where they have been scattered in the cloudy and dark day.
> And I will bring them out from the people,
> And gather them from the countries,
> And will bring them to their own land,
> And feed them upon the mountains of Israel by the rivers
> And in all the inhabited places of the country.
> I will feed them in a good pasture,
> And upon the high mountains of Israel shall their fold be:
> There shall they lie in a good fold,

And in a fat pasture shall they feed upon the mountains of Israel.
I will feed my flock,
And I will cause them to lie down, saith the Lord God (Adonai Jehovah).
I will seek that which was lost,
And bring again that which was driven away,
And will bind up that which was broken,
And will strengthen that which was sick:
But I will destroy the fat and the strong;
I will feed them with judgment."

It is of this judgment with which this book, and indeed the whole Revelation, ends, in the next and final chapter.

But before we come to that we have one more picture in the third constellation of this Sign, which combines the first two in one.

§ 3. ARGO (The Ship).

The Pilgrims safe at Home.

This is the celebrated ship of the Argonauts, of which HOMER sung nearly ten centuries before Christ. Sir Isaac Newton puts the expedition of the Argonauts shortly after the death of Solomon (about 975 B.C.). While Dr. Blair's chronology puts it at 1236 B.C.

Whatever fables have gathered round the story there can be no doubt as to its great antiquity. Some think that the story had its origin in name, as well as in fact, from the *Ark* of Noah and its mysterious journey. All that is clear, when divested of mythic details, is that the sailors in that ship, after all their dangers, and toils, and battles were over, came back victorious to their own shores. The "golden fleece," for which the Argonauts went in search, tells of a treasure that had been *lost.* "Jason," the great captain, tells of Him who recovered it from the *Serpent*, which guarded it with ever-watchful eye, when none else was able to approach it. And thus, through the fables and myths of the Greeks, we can see the light primeval shine; and this light, once seen, lights up this Sign and its constellations, so that their teaching cannot be misunderstood.

ARATUS sings of *Argo*:—

"Stern-foremost hauled; no mark of onward-speeding ship.
Sternward she comes, as vessels do
When sailors turn the helm
On entering harbour: all the oars back-water,
And gliding backward, to an anchor comes."

It tells of that blessed home-coming, when—

"The ransomed of the LORD shall return
And come to Zion with songs,
And everlasting joy upon their heads;
They shall obtain joy and gladness,
And sorrow and sighing shall flee away." (Isa. xxxv. 10.)

It tells of the glorious Jason,* of whom it is asked:

"Art thou not it which hath cut Rahab,
And wounded the dragon?
Art thou not it which hath dried the sea, the waters of the great deep;
That hath made the depths of the sea a way for the ransomed to pass over?
Therefore the redeemed of the LORD shall return,
And come with singing unto Zion," etc. (Isa. li. 9-11.)

---

* The Græco-Judean equivalent of Joshua or Jesus.

"For the LORD hath redeemed Jacob,
And ransomed him from the hand of him that was stronger than he.
Therefore they shall come and sing in the height of Zion,
And shall flow together to the goodness of the LORD."
(Jer. xxxi. 11, 12.)

This is the return of the great emigrant-ship (*Argo*) and all its *company of travellers* (for this is the meaning of the word *Argo*).

In Kircher's Egyptian Planisphere *Argo* is represented by two galleys (as we have two sheepfolds), whose prows are surmounted by rams' heads; and the stern of one of them ends in a fish's tail. One of the two occupies *four segments* of the sphere (from Taurus to Virgo), while the other occupies the four from Leo to Capricorn. *One half* of the southern meridians is occupied with these galleys and their construction and decorations. Astronomers tell us that they carry us back, the one to the period when the Bull opened the year (to which time VIRGIL refers); and the other to the *same* epoch, when the summer solstice was in Leo—"an era greatly antecedent to the Argonautic expedition. How else, they ask, do we account for the one ship having her prow in the first Decan of *Taurus*, and her poop in the last Decan of *Leo?* or for one galley being freighted with the installed *Bull*, and the other with the solstitial *Lion?*" *

These are the words of an astronomer who knows nothing whatever of our interpretation of the heavens which is set forth in this work.

* Jamieson's *Scientific Display*, &c., p. 58.

It will indeed be a large vessel, the true *Argo*, with its *company of travellers*, "a great multitude which no man can number." All this is indicated by the immense size of the Constellation, as well as by the large number of its stars. There are 64 stars in *Argo* (reckoning by the Britannic catalogue); one of the 1st magnitude, six of the 2nd, nine of the 3rd, nine of the 4th, etc. Only a small part of the ship's poop is visible in Britain.

Its brightest star, α (near the keel), is called *Canopus* or *Canobus*, which means *the possession of Him who cometh.* Other star-names, not identified, are—*Sephina, the multitude* or *abundance; Tureis, the possession; Asmidiska, the released who travel; Soheil* (Arabic), *the desired;* and *Subilon, the Branch.*

Is not all this exactly in harmony with the rest of this sign? And is not this what is written in the Book?

> "Therefore, fear thou not, O My servant Jacob, saith the
> Lord;
> Neither be dismayed, O Israel:
> For, lo, I will save thee from afar,
> And thy seed from the land of their captivity;
> And Jacob shall return and be in rest,
> And be quiet, and none shall make him afraid,
> For I am with thee, saith the Lord, to save thee."
> (Jer. xxx. 10, 11.)

> "Lift up thine eyes round about, and see;
> All they gather themselves together, they come to thee;
> Thy sons shall come from far,
> And thy daughters shall be nursed at thy side,
> Then thou shalt see, and flow together,

XXXVIII ARCO. (the Ship)

And thine heart shall fear and be enlarged;
Because the abundance of the sea shall be converted
unto thee. . . .
Who are these that fly as a cloud?
And as doves to their windows?
Surely the isles shall wait for me,
And the SHIPS of Tarshish first, to bring thy sons
from far." (Isa. lx. 4, 5, 8, 9.)

The whole chapter (Isa. lx.) should be read if we wish to understand the great teaching of this Sign, which tells of Messiah's secured possessions, the safe folding of His blood-bought flock, the blessed return of His pilgrims, and their abundant entrance into everlasting rest.

"There is a blessed home
Beyond this land of woe,
Where trials never come,
Nor tears of sorrow flow;
Where faith is lost in sight,
And patient love is crowned,
And everlasting light
Its glory throws around.

O joy, all joys beyond,
To see the Lamb who died,
And count each sacred wound
In hands, and feet, and side;
To give to Him the praise
Of every triumph won,
And sing through endless days
The great things He hath done.

Look up, ye saints of God,
Nor fear to tread below
The path your Saviour trod
Of daily toil and woe;
Wait but a little while
In uncomplaining love,
His own most gracious smile
Shall welcome you above."

CHAPTER IV.

THE SIGN LEO (The Lion).

Messiah's Consummated Triumph.

HERE we come to the end of the circle. We began with Virgo, and we end with Leo. No one who has followed our interpretation can doubt that we have here the solving of the Riddle of the Sphinx. For its *Head* is Virgo and its *Tail* is Leo!

In Leo we reach the end of the Revelation as inspired in the Word of God; and it is the end as written in the heavens.

BAILLY (*Astronomy*) says, "the Zodiac must have been first divided when the sun at the summer solstice was in 1° Virgo, where the woman's head joins the Lion's tail."

As to its antiquity there can be no doubt. JAMIESON says, "the Lion does not seem to have been placed among the Zodiacal symbols, because Hercules was fabled to have slain the Nemean Lion. It would seem, on the contrary, that Hercules, who represented the Sun, was said to have slain the Nemean Lion, because *Leo* was already a Zodiacal sign. Hercules flourished 3,000 years ago, and consequently posterior to the period when the summer solstice accorded with *Leo*" (*Celestial Atlas*, p. 40).

There is no confusion about *this* sign. In the ancient Zodiacs of Egypt (Denderah, Esneh) and India we find the Lion. The same occurs on the

Mithraic monuments, where Leo is *passant*, as he is in Moor's Hindu, and Sir William Jones's Oriental Zodiacs. In Kircher's Zodiacs he is *courrant;* in the Egyptian Zodiacs he is *couchant.*

In the Denderah Zodiac he is treading upon a serpent, as shown in Mr. Edward Cooper's *Egyptian Scenery.*

Its Egyptian name is *Pi Mentekeon*, which means *the pouring out.* This is no pouring out or inundation of the Nile, but it is the pouring out of the cup of Divine wrath on that Old Serpent.

This is the one great truth of the closing chapter of this last Book. It is

THE LION OF THE TRIBE OF JUDAH AROUSED FOR THE RENDING OF THE PREY.

His feet are over the head of *Hydra*, the great Serpent, and just about to descend upon it and crush it.

The three constellations of the Sign complete this final picture:

1. *Hydra*, the old Serpent destroyed.

2. *Crater, the Cup* of Divine wrath poured out upon him.

3. *Corvus*, the Bird of prey devouring him.

The Denderah picture exhibits all four in one. The Lion is presented treading down the Serpent. The Bird of prey is also perched upon it, while below is a plumed female figure holding out *two cups*, answering to *Crater*, the cup of wrath.

The hieroglyphics read *Knem*, and are placed underneath. *Knem* means *who conquers*, or *is conquered*, referring to the victory over the serpent. The woman's name is *Her-ua*, *great enemy*, referring to the great enemy for which her two cups are prepared and intended.

The Hebrew name of the sign is *Arieh*, which means *the Lion*. There are six Hebrew words for Lion,* and this one is used of the Lion *hunting down his prey*.

The Syriac name is *Aryo*, *the rending Lion*, and the Arabic is *Al Asad*; both mean *a lion coming vehemently*, *leaping forth as a flame!*

It is a beautiful constellation of 95 stars, two of which are of the 1st magnitude, two of the 2nd, six of the 3rd, thirteen of the 4th.

The brightest star, α (on the Ecliptic), marks the heart of the Lion (hence sometimes called by the moderns, *Cor Leonis*, *the heart of the Lion*). Its ancient name is *Regulus*, which means *treading under foot*. The next star, β, also of the 1st magnitude (in the tip of the tail), is named *Denebola*, *the Judge* or *Lord who cometh*. The star γ (in the mane) is called *Al Giebha* (Arabic), *the exaltation*. The star δ (on the hinder part of the back) is called *Zosma*, *shining forth*.

* (1) *Gor*, a lion's whelp. (2) *Ciphir*, a young lion when first hunting for himself. (3) *Sachal*, a mature lion in full strength. (4) *Laish*, a fierce lion. (5) *Labia*, a lioness; and (6) *Arieh*, an adult lion, having paired, in search of his prey (Nah. ii. 12; 2 Sam. xvii. 10; Num. xxiii. 24).

XXXIX. LEO (the) LION
CANCER
HYDRA
VIRGO
ECLIPTIC
Regulus
Al Giebha
Zosma
Denebola

Other stars, not identified, are named *Sarcam* (Hebrew), *the joining;* intimating that here is the point where the two ends of the Zodiacal circle have their *joining.* Another star has the name of *Minchir al Asad* (Arabic), *the punishing* or *tearing of the Lion.* Another is *Deneb Aleced, the judge cometh who seizes.* And another is *Al Dafera* (Arabic), *the enemy put down.*

What can be more expressive? What can be more eloquent? All is harmony, and all the names unite in pointing us to what is written of "the Lion of the Tribe of Judah."

And why is Messiah thus called? Because it is applied to Him in Rev. v. 5 in connection with His rising up for judgment: and because the Lion is known to have been always borne upon the standard of Judah, whether in the wilderness (Num. ii.) or in aftertimes.

In Israel's dying blessing the prophetic words foretold of Judah:

> "Thy hand shall be on the neck of thine enemies; . . .
> Judah is a lion's whelp;
> From the prey, my son, thou art gone up.
> He stooped down, he couched as a lion,
> And as an old lion; who shall rouse him up?"
> (Gen. xlix. 8, 9.)

In the prophecy of Balaam (Num. xxiv. 8, 9), we read:

> "He shall eat up the nations his enemies,
> And shall break their bones,
> And pierce them through with his arrows,
> He couched, he lay down as a lion,
> And as a great lion; who shall stir him up?"

The same testimony is borne by the Prophet Amos:

> "Will a lion roar in the forest when he hath no prey?
> Will a young lion cry out of his den, if he hath taken nothing? . . .
> The lion hath roared, who will not fear?"
>
> (Amos iii. 4, 8.)

When "the Lion of the tribe of Judah" is roused up for the rending, the Spirit describes the scene in Isa. xlii. 13:

> "The LORD shall go forth as a mighty man,
> He shall stir up jealousy like a man of war;
> He shall cry, yea, roar;
> He shall prevail against His enemies."

And this is what is meant and included when the Elder says for John's comfort, "the Lion of the Tribe of Judah *hath prevailed*," and hence, is "worthy . . . to receive power, and riches, and wisdom, and strength, and honour, and glory, and blessing" (Rev. v.).

Whether we look, therefore, at the primeval Revelation in the heavens, or at the later Revelation in the Word, the story is one and the same.

And what we see of Leo and his work in both, we find developed and described in the three constellations of the Sign.

§ 1. HYDRA (The Serpent).

The Old Serpent Destroyed.

The time has at length come for the fulfilment of the many prophecies pictured in the heavens: and

XL.
HYDRA (the Serpent)
CRATER (the Cup)
CORVUS (the Raven)
LIBRA
VIRGO
LEO
ECLIPTIC
CENTAURUS
ARGO
Al Gorab
Al Chiba
Al Kes
Al Phard

in its three final constellations we see the consummation of them all in the complete destruction of the Old Serpent, and all his seed, and all his works.

It is the special work of the Messiah, as "the Lion of the tribe of Judah," to trample it under foot.

It is pictured as *the female serpent* (*Hydra*), the mother and author of all evil. *Hydra* has the significant meaning, *he is abhorred!*

It is an immense constellation extending for above 100 degrees from east to west, beneath the Virgin, the Lion, and the Crab. It is composed of 60 stars; one of the 2nd magnitude, three of the 3rd, twelve of the 4th, etc.

The brightest star, α (in the heart of the Serpent), is sometimes called by the moderns *Cor Hydræ* on that account. Its ancient name is *Al Phard* (Arabic), which means *the separated, put away.* Another is called *Al Drian, the abhorred.* Another star is named *Minchar al Sugia, the piercing of the deceiver.*

There can be no doubt as to what is taught by the constellation of Hydra, nor is it necessary to quote the Scriptures concerning the destruction of the Serpent. We pass on to consider the second.

§ 2. CRATER (The Cup).

The Cup of Divine wrath Poured out upon Him.

"God is the Judge.
He putteth down one, and setteth up another,
FOR IN THE HAND OF THE LORD THERE IS A CUP,

And the wine is red; it is full of mixture,
And He poureth out of the same:
But the dregs thereof, all the wicked of the earth shall
wring them out and drink them." (Ps. lxxv. 8.)

"Upon the wicked he shall rain snares,
Fire and brimstone, and a horrible tempest:
THIS SHALL BE THE PORTION OF THEIR CUP." (Ps. xi. 6.)

This is no fabled wine-cup of Bacchus; but it is "The cup of His indignation" (Rev. xiv. 10); "The cup of the wine of the fierceness of his wrath" (Rev. xvi. 19). This is what we see set forth in this constellation. The Cup is wide and deep, and fastened on by the stars to the very body of the writhing serpent. The same stars which are in the foot of the Cup form part of the body of Hydra, and are reckoned as belonging to both constellations.

This Cup has the significant number of *thirteen* stars (the number of Apostacy). The two—*Al Ches* (α), which means *the Cup*, and (β)—determine the bottom of the Cup.

§ 3. CORVUS (The Raven).

The Birds of Prey devouring the Serpent.

Here is the final scene of judgment. We have had *Zeeb, the Wolf;* now we have *Oreb, the Raven. Her-na* is its name in the Denderah Zodiac. *Her,* means *the enemy;* and *Na,* means *breaking up* or *failing.* That is to say, this scene represents *the breaking up* of the enemy.

There are nine stars (the number of *judgment*) in this constellation. The bright star α (in the eye) is called *Al Chibar* (Arabic), *joining together*, from the Hebrew *Chiba* (Num. xxiii. 8), which means *accursed*. This star, then, tells of *the curse inflicted*. The star β (in the right wing) is called *Al Goreb* (Arabic), from Hebrew *Oreb, the Raven*. A third star is named *Minchar al Gorab* (Arabic), and means *the Raven tearing to pieces*.

This brings us to the end. There is nothing beyond this. Nothing remains to be told. We know from the Word of God that—

> "The eye that mocketh at his father,
> And despiseth to obey his mother,
> The ravens of the valley shall pick it out."
> (Prov. xxx. 17.)

We remember how David said to the Giant Goliah—a type of this enemy of God's people—"I will smite thee, and take thy head from thee; and I will give the carcases of the host of the Philistines this day unto the fowls of the air, and to the wild beasts of the earth" (1 Sam. xvii. 46).

When the great day of this judgment comes, an angel standing in the sun will cry "to all the fowls that fly in the midst of heaven, Come, and gather yourselves together unto the supper of the great God; that ye may eat the flesh of kings, and the flesh of captains, and the flesh of mighty men, and the flesh of horses, and of them that sit on them, and the flesh of all men, both free and bond, both small and great" (Rev. xix. 17, 18).

And after these awful words shall be fulfilled, in the closing words of the prophecy of Isaiah, Jehovah foretells us how—

"They shall go forth, and look upon the carcases of the men that have transgressed against Me;
For their worm shall not die,
Neither shall their fire be quenched;
And they shall be an abhorring unto all flesh."

This is the teaching of the whole Sign of LEO! It is all summed up in Jer. xxv. 30–33:—

"Therefore prophesy against them all these words, and say unto them,
The LORD shall roar from on high,
And utter His voice from His holy habitation;
He shall mightily roar upon His habitation;
He shall give a shout, as they that tread the grapes,
Against all the inhabitants of the earth.
A noise shall come even to the ends of the earth;
For the LORD hath a controversy with the nations,
He will plead with all flesh;
He will give them that are wicked to the sword, saith the LORD.
Thus saith the LORD of hosts,
Behold, evil shall go forth from nation to nation,
And a great whirlwind shall be raised up from the coasts of the earth.
And the slain of the LORD shall be at that day from one end of the earth
Even to the other end of the earth;
They shall not be lamented, neither gathered, nor buried;
They shall be dung upon the ground."

Here is the conclusion of the whole matter! Here is the final triumph of the Son of Man in the consummated victory of the Seed of the woman: "Worthy is the Lamb that was slain to receive power, and riches, and wisdom, and strength, and honour, and glory, and blessing" (Rev. v. 12).

"O what a bright and blessed world
  This groaning earth of ours will be,
When from its throne the tempter hurled,
  Shall leave it all, O Lord, to Thee!
But brighter far that world above,
  Where we, as we are known, shall know;
And, in the sweet embrace of love,
  Reign o'er this ransomed earth below.
O blessed Lord! with longing eyes
  That blissful hour we wait to see;
While every worm or leaf that dies
  Tells of the curse, and calls for Thee.
Come, Saviour! Then o'er all below
  Shine brightly from Thy throne above,
Bid heaven and earth Thy glory know,
  And all creation feel Thy love."

Man has ever sought to rob Christ of His glory. He has long since done his best to obliterate His name and His work from the Revelation which had been written in the stars of light. When He humbled Himself, and came as the promised Seed of the woman, men "saw no beauty in Him that they should desire Him." And these were *religious* men. It was religious men, not the common rabble, whom the Old Serpent made use of to wound Him in the heel. The Devil could not touch Him himself; he must use them as his instruments; and it was only *religious* men that could be so used.

It was the "chief priests and scribes," men learned in the Scriptures, whose very knowledge of the Word was used to compass His death amongst the babes at Bethlehem (Matt. ii. 4–6).

It was the same priests and scribes who were used to put Him to death, and give the long-prophesied wound in the heel.

Religion without Christ is enmity against God! Knowledge of the Scriptures where the heart is not subject to Christ, and where Christ is not seen in them, is powerless and lifeless. It is true of the Scriptures, as it will be of the heavenly Jerusalem—"THE LAMB IS THE LIGHT THEREOF" (Rev. xxi. 23).

The Church of Rome has been used of the great enemy to rob the Lamb of God of His promised glory. JEROME, in his Latin translation of the Bible (405 A.D.), wrote "*ipse*," HE, in Gen. iii. 15, as the "bruiser of the serpent's head." And, in spite of the fact that JEROME himself so quotes it in his commentary, and that it is *masculine* in all the other ancient translations of the Bible, Rome has first corrupted JEROME's Vulgate by changing the "e" into "a," and putting "*ipsa*" (she) instead of "*ipse*" (He); then she has so translated this corruption and perpetuated this perversion in various languages! So that in all her versions, in her pictures and statues, in the decree of Pope Pius IX., which promulgated the dogma of the "*immaculate conception of the Virgin Mary*," this lie of the Old Serpent has been foisted on unnumbered thousands of deluded souls, who have thereby been deceived into putting Mary in the place of Jesus; the "co-Redemptress" in the place of the Redeemer; the creature in the place of the Creator; the woman in the place of the woman's Seed;—until the outcome is reached by emblazoning, in huge gilt letters, on the outside of a large church in Rathmines, Dublin, "MARIÆ PECCATORUM REFUGIUM," to Mary the Refuge of Sinners!

So complete has been the success of the subtlety of the Serpent, that he has beguiled thousands of Protestants to unite in circulating these *corrupted versions as the Word of God*, thus giving currency to the Devil's lie. This is done on the plea of expediency, in order that these versions might come to many as Protestant truth instead of Popish error; but thus misleading those who were seeking for light, while confirming Papists in their darkness.

But through all the "wisdom of the Serpent" we can detect his lie. It is very thinly veiled, and the Old Serpent has not succeeded in blinding the eyes which the Spirit of God has opened. True, we see in all Rome's pictures and statues the foot of Mary on the Serpent's head, but the foot is not *coming down*, nor is the head *crushed!* Rather is the woman's foot resting on its head; and the woman herself supported by the Serpent.

The whole system of Mary-anity is thus seen to be the outcome of the Serpent's wisdom in opposition to the true Christ-ianity.

How different are the primeval star-pictures of the heavens. There, the club is lifted up, the foot is coming down, yea, the foot is actually planted upon the enemy, treading the Scorpion under foot.

Rome may corrupt the words of the Book, but she cannot touch the stars of heaven! The Devil himself cannot move them from their places. He may choose and use his servants and agents for corrupting the Scriptures written in the Book, but he cannot change the Revelation of the stars.

There,—no woman's foot is seen upon the Serpent's head! There,—no woman usurps the place of the all-glorious Redeemer!

In *Ophiuchus* we see HIM in dread conflict with the Serpent, and we see HIS foot upon the Scorpion's heart (SCORPIO). We see HIM, the Risen Lamb (ARIES), binding *Cetus*, the great Monster of the Deep; we see HIM in the glorious *Orion*, whose foot is coming down on the enemy's head (*Lepus*); we see HIM in the Lion of the Tribe of Judah (LEO), about to tread down that Old Serpent (*Hydra*) the Devil; we see HIM in the mighty *Hercules*, who has his foot on the head of the *Dragon* (*Draco*), and his up-lifted club about to inflict the long-threatened blow; we see HIM crowned in *Cepheus*, with all His enemies subdued, and His right foot planted upon the Polar Star!

True, we do see a WOMAN in this heavenly and Divine revelation; for there are four women. Two are connected with the REDEEMER, and two with the REDEEMED. The Redeemer is seen in the one (VIRGO) as the "promised Seed"; in the other (*Coma*), He is seen as the child born, the Son given. The Redeemed are represented in one as a captive *chained* (*Andromeda*), with no power to wage conflict with an enemy, but a prey to every foe; in the other (*Cas-*

siopeia), she is *enthroned*, with no necessity for conflict. For with one hand she waves the palm of a victory which another (*Perseus*) has wrought on her behalf, while with her right hand she is preparing and making herself ready for "the marriage of the Lamb."

Thus pure and undefiled is this primeval fountain of Divine truth. Thus harmonious is it with the written Word of God. And He who gave them both to enlighten a dark world which lieth in the power of this wicked one, has filled both with one subject—"The sufferings of Christ and the glory that should follow."

These are set forth by the Holy Spirit in a double sevenfold expansion of the prophetic promise of Gen. iii. 15, giving seven steps in His humiliation and seven in His glorification (Phil. ii. 5–11*).

CHRIST JESUS,

1. Who, being in the form of God, thought it not robbery (a thing to be grasped at and held) to be equal with God;
2. But made Himself of no reputation (Gr. *emptied Himself*),
3. And took upon Him the form of a servant,
4. And was made in the likeness of men:
5. And being found in fashion as a man. He humbled Himself.
6. And became obedient unto death,
7. Even the death of the cross.

* The passage consists really of two members, each of which is arranged as an introversion, where the subject of 1 corresponds to 7; 2 corresponds to 6; etc.

WHEREFORE

1. God also hath highly exalted Him,
2. And given Him a name which is above every name:
3. That at the name of Jesus every knee should bow,
4. Of things in heaven,
5. And things on earth,
6. And things under the earth;
7. And that every tongue shall confess that Jesus Christ is Lord, to the glory of God the Father.

Amen.

"Come, then, and, added to Thy many crowns,
Receive yet one, the crown of all the earth,
Thou who alone art worthy! It was thine
By ancient covenant, ere Nature's birth;
And Thou hast made it Thine by purchase since,
And overpaid its value with Thy blood.
Thy saints proclaim Thee king; and in their hearts
Thy title is engraven with a pen
Dipp'd in the fountain of eternal love.
Thy saints proclaim Thee king; and Thy delay
Gives courage to their foes, who, could they see
The dawn of Thy last advent, long desired,
Would creep into the bowels of the hills,
And flee for safety to the falling rocks."

° ° ° ° ° °

"Come, then, and, added to Thy many crowns,
Receive yet one, as radiant as the rest,
Due to Thy last and most effectual work,
Thy Word fulfilled, the conquest of a world."

(*Cowper.*)

"FOR SIGNS AND FOR SEASONS."

WE have seen the great truths which are taught from the position, and forms, and names of the heavenly bodies. There are also truths to be learnt from their *motions*.

When God created them and set them in the firmament of heaven, He said, in Gen. i. 14—

"Let them be for signs and for seasons."

Here the word "signs" is *othoth* (אֹתֹת, plural of אוֹת, *oth*, from the root אָתָה, *to come*). Hence, *a sign of something* or *some One to come*. In Jer. x. 2 Jehovah says, "And be not dismayed *at the signs* (וּמֵאֹתוֹת) of the heavens, for the heathen are dismayed at them." The word "seasons" does not denote merely what we call the four seasons of the year, but *cycles* of time. It is מוֹעֵד, *appointed time* (from the verb יָעַד, *to point out, appoint*). It occurs three more times in Genesis, each time in connection with the promised Seed—

Gen. xvii. 21, "*At* this *set time* in the next year";

Gen. xviii. 14, "*At the time appointed* I will return"; and

Gen. xxi. 2, "*At the set time* of which God had spoken."

Gen. i. 14 is therefore, "They (the sun, moon, and stars) shall be for signs (things to come) and for cycles (appointed times)."

Here, then, we have a distinct declaration from God, that the heavens contain not only a Revelation concerning *things to come* in the "Signs," but also concerning *appointed times* in the wondrous movements of the sun, and moon, and stars.

The motions of the sun and moon are so arranged that at the end of a given interval of time they return into almost precisely the same position, with regard to each other and to the earth, as they held at the beginning of that interval. "Almost precisely," but not quite precisely. There will be a slight outstanding difference, which will gradually increase in successive intervals, and finally destroy the possibility of the combination recurring, or else lead to combinations of a different character.

Thus the daily difference between the movement of the sun and of the stars leads the sun back very nearly to conjunction with the same star as it was twelve months earlier, and gives us the cycle of the year. The slight difference in the sun's position relative to the stars at the end of the year, finally leads the sun back to the same star at the same time of the year, *viz.*, at the spring equinox, and gives us the great precessional cycle of 25,800 years.

So, too, with eclipses. Since the circumstances of any given eclipse are reproduced almost exactly 18 years and 11 days later, this period is called an *Eclipse Cycle*, to which the ancient astronomers gave the name of *Saros;** and eclipses separated from each other by an exact cycle, and, therefore, corresponding closely in their conditions, are spoken of as being one and the same eclipse. Each *Saros* contains, on the average, about 70 ± eclipses. Of these, on the average, 42± are solar and 28± are lunar. Since the *Saros* is 11 days (or, more correctly, 10·96 days) longer than 18 years, the successive recurrences of each eclipse fall 11 days later in the year each time, and in 33 *Sari* will have travelled on through the year and come round very nearly to the original date.

But as the *Saros* does not reproduce the conditions of an eclipse with absolute exactness, and as the difference increases with every successive return, a time comes when the return of the *Saros* fails to bring about an eclipse at all. If the eclipse be a solar one before this takes place, a new eclipse begins to form a month later in the year than the old one, and becomes the first eclipse of a new series.

This is the history of one such eclipse: On May 15 (Julian), 850 A.D., there was a (new) eclipse of the sun, and it occurred as a *partial* eclipse. On August 20 (Julian), 1012 A.D., this new eclipse became *total.* From that time it has been an *annular* eclipse, the latitude of the central shadow gradually shifting southward from the north, until on December 17 (Julian), 1210, it had reached N. Lat. 24°. It turned northward again after 1210, until March 14 (Julian), 1355, when it fell in N. Lat. 43°. Then it turned south,

---

* General Vallancey spells *Saros* שׁערוֹין, which amounts to 666 by Gematria! *Viz.*, ש = 300 + ע = 70 + ר = 200 + ו = 6 + ץ = 90 = 666.

M 2

and has moved steadily in that direction, until on March 18 (Greg.), 1950, its last appearance as an annular eclipse will take place. On May 22 (Greg.), 2058, it will fall so far from the node that a new eclipse will follow it on June 21. . It will make three more appearances as an ever-diminishing partial eclipse, and be last seen on June 24 (Greg.), 2112. Its total life-history, therefore, will have been 1,262 years and 36 days, and will have occupied 70 *Sari*.

In the above life-history of an eclipse * there is not the slightest difficulty as to its identification. The *Saros* shows no break, and no interruption; nor does the character of the eclipse suffer any abrupt change. The district over which it is visible moves in a slow and orderly fashion from occurrence to occurrence over the earth's surface.

Now the important point is this, that if we take the prophetic reckoning of 360 days to the year, we have the following significant Biblical numbers :—

In the first place, we already have the 70± *Sari* divided into two portions of 33 + 37.

A perfect cycle is accomplished in 33 *Sari*, or 595 years, when the eclipse, by a series of unbroken

---

* These facts are kindly supplied by Mr. E. W. Maunder, of the Royal Observatory, Greenwich, who gives another example, as follows :—

In A.D. 586 there were two solar eclipses : on June 22 (Julian) the old and dying eclipse, and on July 22 (Julian) another (the new one). A *Saros* (viz. 18 years and 11 days) earlier *there was only one*, viz. on June 11 (Julian), A.D. 568, there being no eclipse on July 11 of that year.

The last appearance of this new eclipse, which first appeared on July 22, 586, was on August 28 (Greg.), 1848, so that it had a life history of 70 *Sari*, amounting to 1,262 years 36 days (after the Julian dates have been corrected to correspond to the Gregorian). Thus the eclipse that died, so to speak, on August 28 (Greg.), 1848, first appeared on July 22 (Julian) in A.D. 586. See an important article on Eclipses by Mr. E. W. Maunder in *Knowledge*, for October 1893, where other *life-histories* of eclipses are given, and the whole subject of eclipses clearly explained.

Sari, has accomplished a passage through the year of 360 days; or, if we reckon only the whole numbers, *i.e.*, the 18 completed years, we have for the 33 *Sari* the period of 594 years, while the remaining portion of 37 *Sari* makes 666 years (37×18); and the whole 70± *Sari* makes 1,260 years (594+666).*

We have then the following figures:—

18 × 33	=	594 years.
18 × 37	=	666 years.
18 × 70	=	1260 years.

Independently of this, we also know that 1,260 years is a soli-lunar cycle, so exact that its epact, or difference, is only 6 hours!

There must, therefore, be something significant in these numbers, *e.g.*, 70; in the number 1,260, with its divisions, not into two equal parts, but into 594 and 666; as also in its double, 2,520.

There must be something to be learned in the occurrence and repetition of these heavenly cycles, which for nearly 6,000 years have been constantly repeated in the heavens, especially when we find these same numbers very prominently presented in the Word of God in connection with the fulfilment of prophecy.

We have the great "seven times" (2,520) connected with the duration of Israel's punishment, and of the Gentiles' power. We have in Daniel and the

* The relations between 595 years and 1,262 years 36 days, are the same as the relations between 594 years and 1,260 years. The difference of the 2 years 36 days is due to the excess of 10·96 days over the 18 completed years in each *Saros*.

Apocalypse the half of this great period presented as "days" (1,260), as "months" (42), and as "times," or years ($3\frac{1}{2}$).

Futurists believe that these "days" and "months," etc., *interpret for us* the purposes and counsels of God as connected with "the time of the end," and as meaning literal "days" and "months," etc.

Historicists take these terms, and themselves *interpret the numbers*, in the sense of a "day" being put for a *year*, and they believe that these "1,260 days" will be fulfilled as 1,260 *years*.

One party boldly and ungraciously charges the other with teaching "*The Fallacies of Futurism*"; while the other might well retort with a reference to the *Heresies of Historicism*.

But is there any necessity for the existence of two hostile camps? Is it not possible that there may be what we may call a *long* fulfilment in years? And is it not more than probable that in the time of the end, the crisis, there will be also a *short* and literal fulfilment in days?

We firmly believe that there will be this literal and *short* fulfilment. We believe that when God says "days," He means *days;* and that when He says "42 months," He means *months*, and not 1,260 years. In all the passages referred to by historicists in support of what is called "the year-day theory," the Holy Spirit uses these words "days" and "years" in the sense of days and years. In the two particular instances of Israel's wanderings (Num. xiv. 34), and Ezekiel's prophesying (Ezek. iv. 6), He chooses to take the *number* of days as denoting the *same number* of years; but He does not tell us that we are to do the same in other cases! He only asserts His sovereignty by thus acting, while we only show our presumption in taking His sovereign act as a general principle.

But while fully believing in the *short* fulfilment, we are quite prepared to admit that there may be a *long* fulfilment *as well;* and that, owing to the wondrous harmony, and marvellous correspondence, and infinite wisdom of all the works and ways of God, there may be a fulfilment, or rather a "filment," if we may coin the word, in years, which will be only a foreshadowing of the literal *ful*-filment afterwards to take place in *days*.

If historicists will allow us this liberty as to *interpretation*, and permit us to believe that God means what He says, we will give them some remarkable evidence in support of their views, by way of *application.* In other words, if they will allow us to *interpret* "days" as meaning days, we will gladly allow them, and be at one with them, in *applying* them to years. So that while we believe the *interpretation* to mean "days," and to teach a *short* fulfilment at the time of the end, we will thankfully admit an *application* which shall take these days as foreshowing a *long* fulfilment in years.

In *applying*, then, these significant numbers (42, 70, 594, 666, 1,260, and 2,520) to years, from what point or date shall we begin to reckon the "*times of the Gentiles*" (Luke xxi. 24)? That there are such definite "times" the words of the Lord Jesus show, when He says, "Jerusalem shall be trodden down of the Gentiles, until the times of the Gentiles be fulfilled." (Luke xxi. 24). That there are "seven times" of Gentile dominion is more than intimated by the symbolic episode in the life of Nebuchadnezzar as recorded in Dan. iv.; and that there are "seven times" of Israel's punishment is clearly stated in Lev. xxvi. 18. "Seven times," according to the Historicist school of interpreters, are equal to 2,520 years.

Instead of asking where they begin, let us first note the fact that it is *duration* which is emphasised in the Scriptures rather than *chronology;* and look at the duration of these years independently of, and before we attempt to fix, their beginning and ending.

In Daniel ii. and vii. it is shown first to Nebuchadnezzar in a "dream," and afterwards to God's servant the prophet in a "vision," that Israel was to be put on one side and become "Lo-Ammi" (*not My people*), while government was to be put into the hands of the Gentiles. Jerusalem was the central point of both these great and solemn facts. That is to say, during 2,520 years, while Jerusalem should remain in the power of the Gentiles, Israel could be "no more a nation" in possession of their land and city.

We know, as a matter of fact, that to-day Jerusalem is in the hands of the Turks, and that it is still "trodden down of the Gentiles."

If we ask how long it shall continue to be "trodden down"? how long it will be before Israel shall again possess their city and their land?—the answer brings us at once to the heart of our subject.

In seeking to determine both duration and chronology, it is necessary to plant our feet on sure ground. To do this, let us take a point on which all are agreed.

There is *one* date which is universally accepted; and concerning which the evidence is unquestioned.

ABU OBEIDA, the Mahommedan General, laid siege to Jerusalem towards the close of 636 A.D. The city was then occupied by the Romans, who held out for four months. When they capitulated, the Patriarch SOPHRONIUS obtained a clause in the treaty giving security to the inhabitants, and requiring the ratification of OMAR himself. OMAR, who had therefore to be sent for, arrived some six months afterwards, and the delay caused the actual delivering up of the city to take place early in the autumn of A.D. 637.*

The year A.D. 636–7 is therefore the accepted date of the passing over of Jerusalem from the Romans to the Turks.

OMAR seems to have stayed in the city only about ten days, during which he must have given his instructions for the erection of the Mosque on the site of the Temple. This Mosque, therefore, stands as the sign and the symbol of the treading down of Jerusalem, and while it remains, those times of treading down cannot be considered as fulfilled.

How steady was Israel's decadence from Nebuchadnezzar to OMAR! Nothing would exceed that darkest moment in Israel's history, when Israel was

* This is the date which concerns only the *City of Jerusalem*. The Romans were not completely driven out from *the land* until Cæsarea had fallen in 638, when the conquest was finally completed. See Gibbon's *Decline and Fall*.

well nigh obliterated in the mighty struggles of her enemies who fought over her inheritance. Thus OMAR becomes the great central point of the 2,520 years, whether reckoned as *Lunar*, *Zodiacal*, or *Solar*, dividing them equally into two portions of 1,260 years.*

Having thus fixed the central date, which already points forward to the end, let us go back and find the starting point, that we may the better understand the end.

When Daniel was explaining to Nebuchadnezzar his mysterious dream, he said, "Thou art this head

---

* This date 636–7 is a great and important central date, whether we reckon backwards or forwards; whether we reckon them as *Lunar*, *Zodiacal* (360 days), or *Solar* (365 days) years.

If we take *Lunar years* (=1222½ Solar)—

reckoning *backward*, we get to 587 B.C., the very date of the destruction of the Temple by Nebuchadnezzar.

reckoning *forward*, we get to 1860 A.D., the very date of the European intervention in the Lebanon, which has brought the Eastern Question into its present prominent position.

If we take *Zodiacal years* (=1242 Solar)—

reckoning *backward* we get to 608 B.C., the date of the battle of Carchemish (2 Chron. xxxv. 20), when Babylon completed the conquest of Assyria, and became supreme; utterly shattering all the hope which Israel had in Egypt.

reckoning *forward* brings us to 1879 A.D., when, by the Treaty of Berlin, Ottoman power received a blow from which it has never recovered, and which has prepared the way for its extinction.

If we take *Solar years*, then—

reckoning *backward*, we get to B.C. 624 (A.M. 3376), the beginning of the Babylonian kingdom, the "head of gold."

reckoning *forward* we get to 1896–7 A.D., which is yet future.

These reckonings in their *beginnings* and *endings* form an *introversion*, or *Epanodos*, thus:—

587
  608 } B.C. dates increasing.
    624
    1860
  1879 } A.D. dates increasing.
1896–7

The *Solar* reckonings are the more important dates; the *Lunar* are next in significance; while *Zodiacal* reckonings furnish us with dates which, to say the least, fit neatly into their places.

of gold"! (Dan. ii. 38). This moment is popularly, but erroneously, supposed to mark the commencement of the Babylonian kingdom—the first of these four great Gentile powers.

But Daniel spoke of what ALREADY existed, and was *explaining the then* condition of things. He said, "God *hath* given thee a kingdom, power, and strength, and glory" (Dan. ii. 37). The kingdom of Babylon had already been in existence for more than thirty years, for its king had destroyed Jerusalem and burnt the Temple with fire, and brought away many captives, amongst whom was Daniel and his companions. The opening words of the book make this very clear.

The monumental history of Babylon, as now dug up, shows that before this it had been sometimes tributary to, and sometimes almost independent of, Assyria. In A.M. 3352, after a severe struggle with Assurbanipal, the Assyrian king, Babylon was once more subdued, and its king setting fire to his palace perished in the flames. After that there was peace for twenty-two years, during which time Kandalanu governed Babylon in succession to Sumas-sum-ukin, a son of Assurbanipal.

In A.M. 3375 (*i.e.* B.C. 627),* another revolt broke out, and the Assyrian king sent a general of great ability to quell it. His name was Nabu-pal-user (which means *Nebo protects his son*). He put down

* These dates are those furnished by the Monuments, as given by Dr. Budge, of the British Museum, in his *Babylonian Life and History*, R.T.S., 1885. They also agree with the dates dug up by Sir Henry Rawlinson in 1862, consisting of fragments of seven copies of the famous "Eponym Canon of Assyria," by which the Assyrian chronology has been definitely settled. Before this, historians had to be content with inferences and conjectures.

the rebellion with so much skill that Assurbanipal made him governor of Babylon. He shortly afterwards, in A.M. 3376, himself rebelled, and made himself King of Babylon. Associating with him his son Nebuchadnezzar, they commenced a campaign against Assurbanipal, which ended in the fall of Nineveh and the complete subjugation of Assyria. The kingdom of Babylon, thus commencing in B.C. 625,* became the first great Gentile kingdom as foretold in Daniel.

There is practically no question, now, as to this date.

The actual *duration* of the three kingdoms that followed—Babylon, Medo-Persia, and Greece, may not perhaps be so accurately determined. Their total duration is known, because it is fixed by a known date at the other end, but it might introduce controversial matter if we attempted to assign to them their exact relative duration. Probably they were, roughly:—Babylon about 90 years; Medo-Persia about 200 years; Greece about 304 years.

We believe these to be fairly proportionate,† but whether they are or not, their total amount must have been 594 years, because the undisputed date of the battle of ACTIUM, by which Augustus became the head of the Roman Empire, was September B.C. 31.

* In adjusting the A.M. and B.C. dates, the latter are always apparently one year in advance of the former, because B.C. 4000 was A.M. 1, and B.C. 3999 was A.M. 2. Hence A.M. 3376 is not B.C. 624, but it is B.C. 625.

† Cyrus took Babylon, according to the Monuments, in the 17th year of Nabonidus, B.C. 539. 1 Maccabees i. begins the first of Alexander from the death of Darius Codomannus in A.M. 3672. This would slightly vary the above distribution of the years of separate duration.

From this date Jerusalem passed permanently under the power of Rome until the Mahommedan conquest in A.D. 636–7.

We have, therefore, *three fixed dates*, and these decide for us the *duration* of the intervening periods; dividing them into the two great Eclipse Cycles of 594 years and 666 years!

Jerusalem under the Gentiles.

		Fixed Dates.	Duration of Years.
Babylon (the 1st Kingdom) commenced	B.C.	625	
Battle of Actium, ending the possession of the 3rd Kingdom - - -	B.C.	31	
Duration of the three Kingdoms, Babylon, Medo-Persia, and Greece, together (1st Eclipse Cycle) - - - -			594
Rome (the 4th Kingdom) became the possessor of Jerusalem - -	B.C.	31	
Mahommedan conquest of Jerusalem, ending the possession of Rome -	A.D.	636	
Duration of Rome's possession of Jerusalem (2nd Eclipse Cycle) -			666*
FIRST HALF OF "THE TIMES OF THE GENTILES" - - - - -			1260
Date of Mahommedan conquest of Jerusalem - - - - -	A.D.	636–7	
SECOND HALF OF "THE TIMES OF THE GENTILES" and *Duration* of Mahommedan possession of Jerusalem -		1260	1260
End and "fulness" of "the times of the Gentiles" - - -	A.D.	1896–7	2520

* In passing from B.C. dates to A.D. dates, *one year must always be deducted*, *e.g.*, from B.C. 2 to A.D. 2 is only *three* years, not four! Thus—

From Jan. 1 B.C. 2 to Jan. 1 B.C. 1 is *one* year
,, ,, B.C. 1 to ,, A.D. 1 is *one* year
,, ,, A.D. 1 to ,, A.D. 2 is *one* year
} making only *three* years.

Hence, B.C. 31 to A.D. 636 is 666 years, not 667.

From this it appears that 1896–7 A.D. would mark the fulfilment, and complete the duration, of "times of the Gentiles."

The above reckoning has the following advantages over all previous historicist interpretations:—

1. *Controverted* dates are excluded.

2. The *whole* period of 2520 years is dealt with, instead of only the latter half (1260), as is usually the case.

3. It confines these "times" to the one place where the Lord Himself put them, *viz.*, "JERUSALEM." He said, "Jerusalem shall be trodden down of the Gentiles, till the times of the Gentiles be fulfilled."

These "times," therefore, are confined to Jerusalem. This "treading down" is confined to Jerusalem. It is not the city of Rome that is to be trodden down for 1260 years. Why, then, should these "times" be separated from what is characteristic of their *duration*, and applied to Rome, papal or imperial? Why should historicists search for some act of emperors or popes in the early part of the seventh century in order to add it to 1260, so as to find some terminal date in or near our own times! *

We claim that the Lord Himself has joined these "times of the Gentiles" with the city of "Jerusalem," and we say, "What, therefore, God hath joined together, let not man put asunder" (Matt. xix. 6).

* While the *premisses* of the Historicist school are thus strengthened, their *conclusions* are shown to be erroneous.

When Jesus spoke of this *treading down*, it looks as though it were then still future; for He said, "Jerusalem *shall be* trodden down," etc. The occupation of Jerusalem by Babylon, Medo-Persia, Greece, and Rome, was for purposes of *government* rather than for a wanton treading down. Government on the earth was committed unto them. But when Jerusalem passed from the government of the Roman Empire into the hands of the Turks, it could then be said, in a very special sense, to be "trodden down." For of any government worthy of the name there has been none; and of desolation and desecration the city has been full. Under the feeble rule of the Turks, all the Gentiles seem to have combined in laying waste the holy city.

Though Jews are returning thither in ever-increasing numbers, they are only strangers there. They have as yet no independent position, nor can they make any treaties. But when these "times" shall end, it means that they will have a position of sufficient independence to be able to make a treaty or league with the coming Prince (Dan. ix. 27); and then the course of events will bring on another treading down of 1260 literal "days," which will thus have had a fore-shadowing fulfilment in years! This is written in Rev. xi. 2. And to save us from any misunderstanding, the time is given, not in days, but in "*months*."

The angel, after directing John to measure the Temple of God and the altar, adds, "but the court which is without the Temple leave out, and measure it not; for it is given unto the Gentiles; and the holy city shall they tread under foot forty and two months."

This refers to a future treading down, which will be limited to the brief period of "forty two" literal "months," during the time of the coming Prince; and "in the midst" of the last week, when he shall break His covenant with the Jews,* set up the "abomination of desolation" (Dan ix. 27; which is still future in Matt. xxiv. 15), and "tread down the holy city."

We now desire to specially emphasize the fact that all these dates, and their termination in a rapidly approaching fulfilment, refer ONLY TO JERUSALEM, AND THE GENTILES, AND THE JEWS! They refer only to the end of the Gentile possession of Jerusalem, and to the settlement of the Jews in their own city and land.

These "times and seasons" have nothing whatever to do with "the Church of God" (1 Thess. v. 1). The mystical Body of Christ, whenever its members are complete, "will be taken up to meet the Lord—the Head of the Body—in the air, so to be ever with the Lord" (1 Thess. iv. 15–17). This glorious event has nothing to do with any earthly sign or circumstance, so far as the members of this mystical Body are concerned.

---

* And cause sacrifice and oblation to cease (Dan. ix. 27). We know that is referred, by historicists, to the Messiah. But they are not entitled to so interpret this passage unless they take with it viii. 11, xi. 31, and xii. 11, where the same event is distinctly referred to, and is spoken, not of Christ, but of Antichrist.

Therefore we are not dealing here with the coming of the Lord; either for His saints, or with them. We are not referring to what is commonly and erroneously called "the end of the world." We are merely pointing out that the end of Gentile dominion *over Jerusalem* is drawing near! And we cannot close our eyes to the marvellous manner in which the veil is being removed from Jewish hearts: to the change which has come over the Jewish nation in its attitude towards Christ and Christianity, chiefly, under God, through the unparalleled circulation of more than a quarter of a million copies of a new translation of the New Testament into Hebrew, by the late Isaac Salkinson, published by the Trinitarian Bible Society, and freely distributed by the Mildmay Mission to the Jews: to the Palestine literature which has sprung up amongst the Jews in recent years: to the persecutions in various countries which are stirring their nest, and setting the nation in motion: to the organized emigration to Argentina, which its promoters avowedly speak of as "a nursery ground for Palestine" (*Daily Graphic*, March 10th, 1892): to the railways completed and in course of construction in the Holy Land: to the numerous Societies and their branches which have permeated the whole nation, which, while having various names, have only one object—"the colonisation of Palestine."

When we put these events side by side with the teaching of the heavens as to the "cycles" or appointed times, we are merely showing how wonderfully they agree with what is written in the Book, and witnessed to by great and uncontested historic dates.

Nor are we absolutely naming a definite year or day even for these Palestine events. After all, they can be only approximate, for man has so misused every gift that God has ever given him, that even with such wondrous heavenly time-keepers he cannot really tell you what year it is! And, besides this loss of reckoning, there is confusion as to the commencement of the A.D. era, which makes absolute accuracy between the A.M., B.C., and A.D. dates impossible.

Added to this, there is another point to be borne in mind, *viz.*, that when the "times of the Gentiles" shall end, Jewish independence need *not be either immediate or complete!*

For when Nebuchadnezzar began his kingdom of Babylon in A.M. 3376 (B.C. 625), the Jews, though in their land and city, were not independent. Nebuchadnezzar went to and fro to Jerusalem, and put down and set up whom he would; and it was not till some thirty years afterwards that he destroyed the City and Temple and made the people captives.

So, likewise, in the time of the end, there may be an *epanodos*. There may be a similar period of possession without independence, a quasi-independence guaranteed by the Great Powers; and, for ought we know, it may be that, in order to gain *complete* independence, they may ultimately make that fatal league with the coming Prince.

So that while we name the dates 1896–7 as being significant, we are not "fixing dates" in the ordinary sense of the term, but merely pointing out some of "the signs of the times," concerning which we ought not to be ignorant.

The *true interpretation* will in any case still remain, and will surely be literally fulfilled in its own time. The Word of God will be vindicated; its prophetic truth will be verified; God Himself will be glorified; and His people saved with an everlasting salvation.

Meanwhile the members of His Body will "wait for His Son from heaven, whom He raised from the dead, even Jesus, which delivered us from the wrath to come" (1 Thess. i. 10). They will live "looking for that blessed hope, and the glorious appearing of the great God and our Saviour Jesus Christ, who gave Himself for us, that He might redeem us from all iniquity, and purify unto Himself a peculiar people (R.V., a people for His own possession) zealous of good works" (Titus ii. 13, 14). They will "look for the Saviour, the Lord Jesus Christ," from heaven, believing that there is no hope either for "the Jew, the Gentile, or the Church of God," or for a groaning creation, until "the times of refreshing shall come from the presence of the Lord; and He shall send Jesus Christ, whom the heaven must receive until the times of restitution of all things, which God

N 2

hath spoken by the mouth of ALL HIS HOLY PROPHETS SINCE THE WORLD BEGAN" (Acts iii. 19–21).

"The world is sick, and yet not unto death;
There is for it a day of health in store;
From lips of love there comes the healing breath,—
The breath of Him who all its sickness bore,
And bids it rise to strength and beauty evermore.

Evil still reigns; and deep within we feel
The fever, and the palsy, and the pain
Of life's perpetual heartaches, that reveal
The rooted poison, which, from heart and brain,
We labour to extract, but labour all in vain.

Our skill avails not; ages come and go,
Yet bring with them no respite and no cure;
The hidden wound, the sigh of pent-up woe,
The sting we smother, but must still endure,
The worthless remedies which no relief procure,—

All these cry out for something more divine,
Which the worst woes of earth may not withstand;
Medicine that cannot fail—the oil and wine,
The balm and myrrh, growth of no earthly land,
And the all-skilful touch of the great Healer's hand.

Man needs a prophet: Heavenly Prophet, speak,
And teach him what he is too proud to hear.
Man needs a priest: True Priest, Thy silence break,
And speak the words of pardon in his ear.
Man needs a king: O King, at length in peace appear."

APPENDIX.

APPENDIX.

Note on the Sign *LIBRA.*

On page 47 we called attention to the point that in all probability the Sign Libra was a very ancient corruption.

The ancient Akkadian name for the *seventh* month, which was the month when the sun was in the Sign now called Libra, was *Tul-ku*, which means *the sacred mound*, or *altar*. The Akkadian name for this Sign was *Bir*, which means *the Light*, hence, the Lamp with its light, or the Altar with its fire.

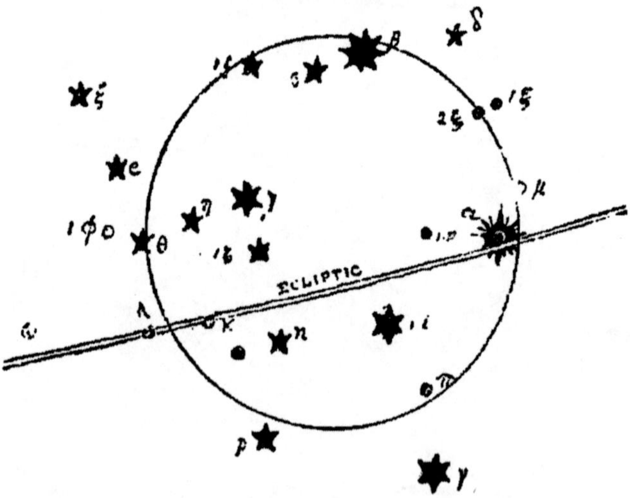

Fig. 1.—The *Circular Altar*, in the Sign now called *Libra*.

Its most ancient form was a circular altar.* In Figure I we have reproduced this,† and it will be at once seen that we have the original of the disc now preserved in the *two circular scales* which form the Sign of LIBRA.

The next stage of the corruption is shown in the Akkadian name of *Scorpio* (the Scorpion)—the Sign immediately to the left of the Altar. It was called *Gir-tab*, which means *the Seizer and Stinger*, and the next Figure (2), taken from an Euphratean boundary stone,‡ shows the two Signs combined, for the Scorpion is stretching out its claws in order to *seize* the *Lamp* or *Altar*.

FIG. 2.—The *Scorpion* and the *Lamp*. (From an Euphratean Boundary Stone.)

---

* See ARATOS, line 440.

† As proved by Mr. Robt. Brown, junr., in his *Remarks on the Euphratean Astronomical Names of the Signs of the Zodiac* (p. 16).

‡ By the kind permission of Mr. Robt. Brown, junr., *The Celestial Equator of Aratos*, p. 466.

Thus the meaning of its name is exemplified. It is called the *Seizer and Stinger.* And just as in the constellation above it, the Serpent is struggling with the man, while at the same time it is stretching out its neck to seize the crown,* so here the Scorpion, while trying to *sting* the same man in the heel, is stretching out its claws to *seize* the altar.

A seal on a contract, nearly 700 B.C., shows this Circular Altar actually in the grasp of the Scorpion.

FIG. 3.—*Scorpion* and *Lamp.* (From an Euphratean Seal.)

Figure 3 is a picture of this Euphratean Seal, preserved on a contract made on the 8th day of the month *Tisri, i.e.*, this same *seventh* month! †

This then is the next stage. But Mr. Robert Brown, junr., observes, "The *Circle* or other representation of an *Altar* not unnaturally disappeared as

* See this shown on the cover of this book.

† Menant, *Empreintes de Cachets Assyro-Chaldéens*, 9. "Sur un contrat daté du 8 Tisri, de l'année de Bin-takkil-ani, 690 ou 645 avant J.C."

the use of the Sign advanced westward; whether by sea, or across Asia Minor, or both, and the *Chelai* alone remained when the shores of the Ægean were reached." *

This is quite true, for the Greek name for the Sign was *Chelai*, which means simply *the Claws*. And thus the Scorpion monopolised two Signs; its body one, and its claws the other. This led to the mistake of SERVIUS, the intelligent commentator on VIRGIL,† that "the Chaldean Zodiac consisted of but eleven constellations." We now know that there were twelve Signs, and the mistake is thus explained.

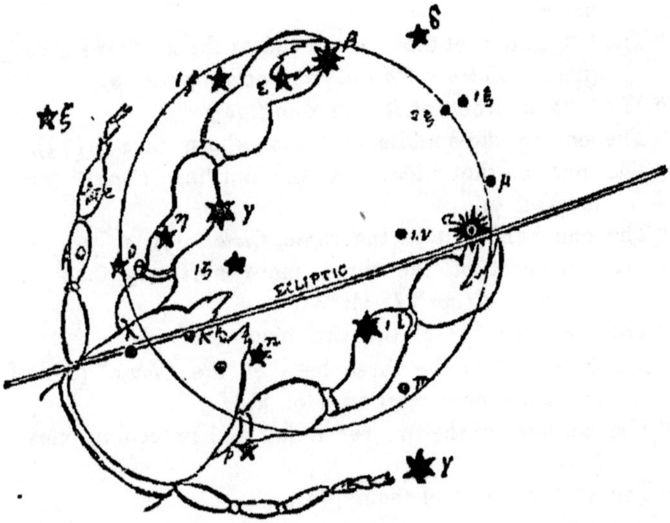

FIG. 4.—The Constellation of "the Claws." Formerly the *Circular Altar*, now *Libra*.

* *Researches on the Euphratean Astronomical Names of the Signs of the Zodiac*, p. 17.

† In *Georgica*, i. 33.

Mr. Brown quotes ACHILLES TATIUS, about 475 A.D., in a Fragment on the *Phainomena*, who says, τὰς χηλὰς τὰς καλόυμένας ὑπ 'Αιγυπτίων Ζυγὸν.*

ARATUS says that "some few stars of the *Claws* are in the (Celestial) Equator." And PTOLEMY describes the stars, now reckoned in LIBRA, as being in what he calls "The Constellation of the Claws." We have reproduced them so that his description of them may be readily traced. He speaks of—

"The bright one of those at the end of the southern *Claw*." (It is named *Zuben el Genubi* and now marked α).
"The one more northerly than it, and dimmer" (now named μ).
"The bright one of those at the end of the northern *Claw*" (named *Zuben el Chemali*, and now marked β).
"The one in front of it and dim" (δ).
"The one in the middle of the southern *Claw*" (1 i).
"The one in the middle of the northern *Claw*" (now marked γ).
"The one behind it in the same *Claw*" (η).
"The foremost of the three more northerly than the northern *Claws*" (1 f).
"The southern one of the two hindmost" (ι).
"The hindmost of the three between the *Claws*" (one of the stars now marked κ or λ).
"The northern of the two remaining and preceding ones" (ζ).
"The southern one of them" (n).

This is how the stars formerly in the Sign of the (Circular) ALTAR. came to be reckoned in *the Claws*

---

* AP PETAVIUS, *Uranologion*, 168, "*The claws, called by the Egyptians Zugon*," i.e., *the yoke* that joins any two things together.

of the Scorpion; and this is how the circular scales of LIBRA came to be substituted for the ancient *Circular* ALTAR.

This corruption of the primitive teaching of the ALTAR, shows how the enemy attempted to *seize* on the Atonement, bring in "the way of Cain," and substitute *human merit* for the atoning sacrifice of Christ; thus perverting the truth at its fountain head. Just as in Gen. iii. we have the woman's promised Seed in conflict with the Enemy, so in Gen. iv. we see the Scorpion's claws—"the way of Cain" in conflict with "the way of God."

There can be but little doubt, therefore, that the first Sign of the Zodiac was VIRGO, the second was the ALTAR, and the third was the SCORPION. The lesson which they teach is clear: The Seed of the woman (VIRGO), who was to come as a child, should be a sacrifice (the ALTAR) for the sins of His people; endure a great conflict with the enemy (SCORPIO), in which He should be wounded in the heel; but should in the end crush and tread the enemy under foot.

CETUS
ERIDANUS
ORION
TAURUS
ARIES
PISCES
ANDROMEDA
PERSEUS
AURIGA
ECLIPTIC
AQUARIUS
PEGASUS
CASSIOPEIA
CEPHEUS
Al Ruccaba
Polar Star
URSA MINOR
PICIS AUSTRALIS
CYGNUS
Deneb
DRACO
DELPHINUS
SAGITTA
LYRA
HERCULES
CAPRICORNUS
THE EQUATOR OF THE HEAVENS
Al Tair
AQUILA
CORONA BOREALIS
Ras al Hague
SERPENS
OPHIUCHUS
SAGITTARIUS
THE ECLIPTIC OR THE SUN

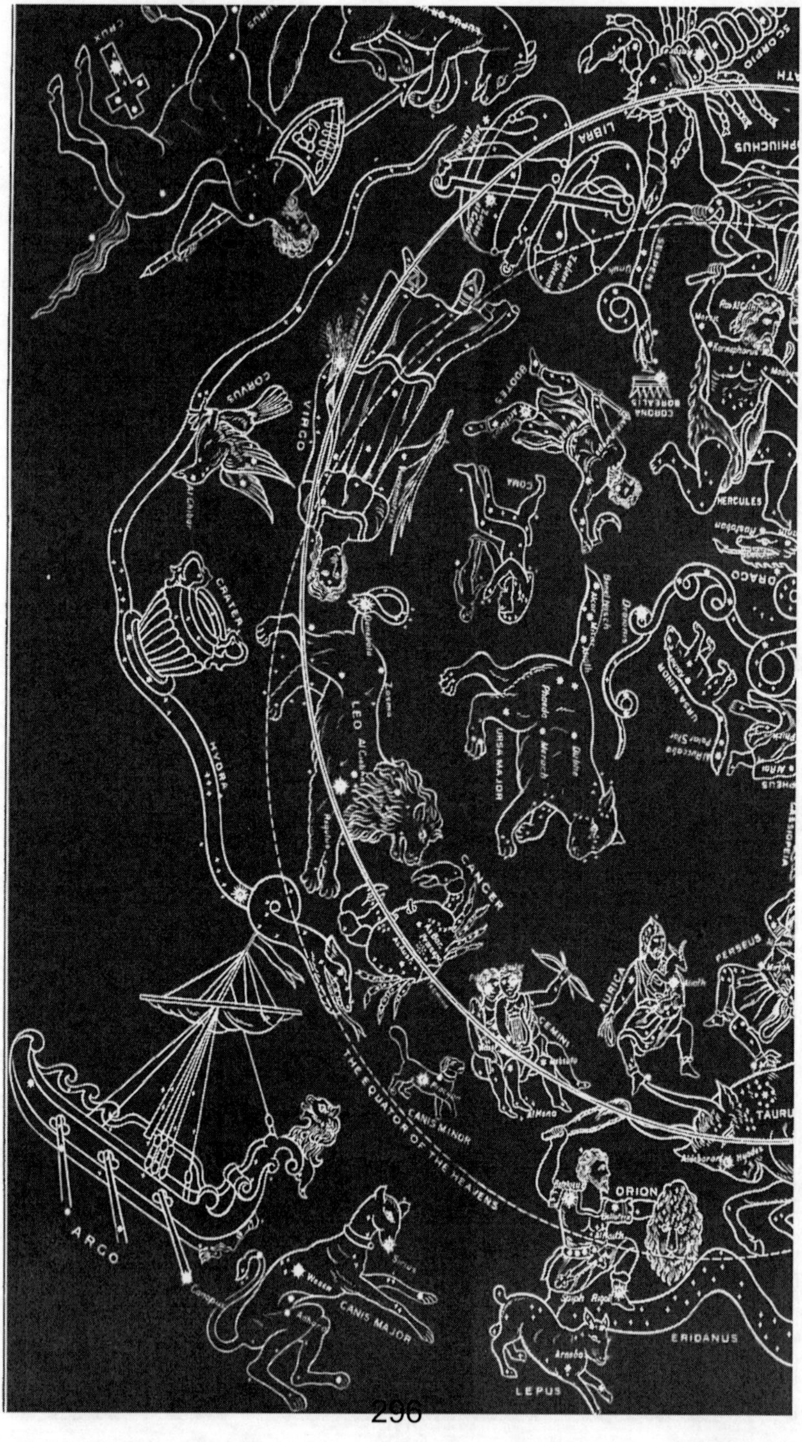
CRUX
CORVUS
VIRGO
CRATER
HYDRA
LEO
URSA MAJOR
BOOTES
COMA
CORONA BOREALIS
HERCULES
LIBRA
SCORPIO
URSA MINOR
CANCER
GEMINI
AURIGA
PERSEUS
ORION
ERIDANUS
LEPUS
CANIS MAJOR
CANIS MINOR
ARGO
THE EQUATOR OF THE HEAVENS

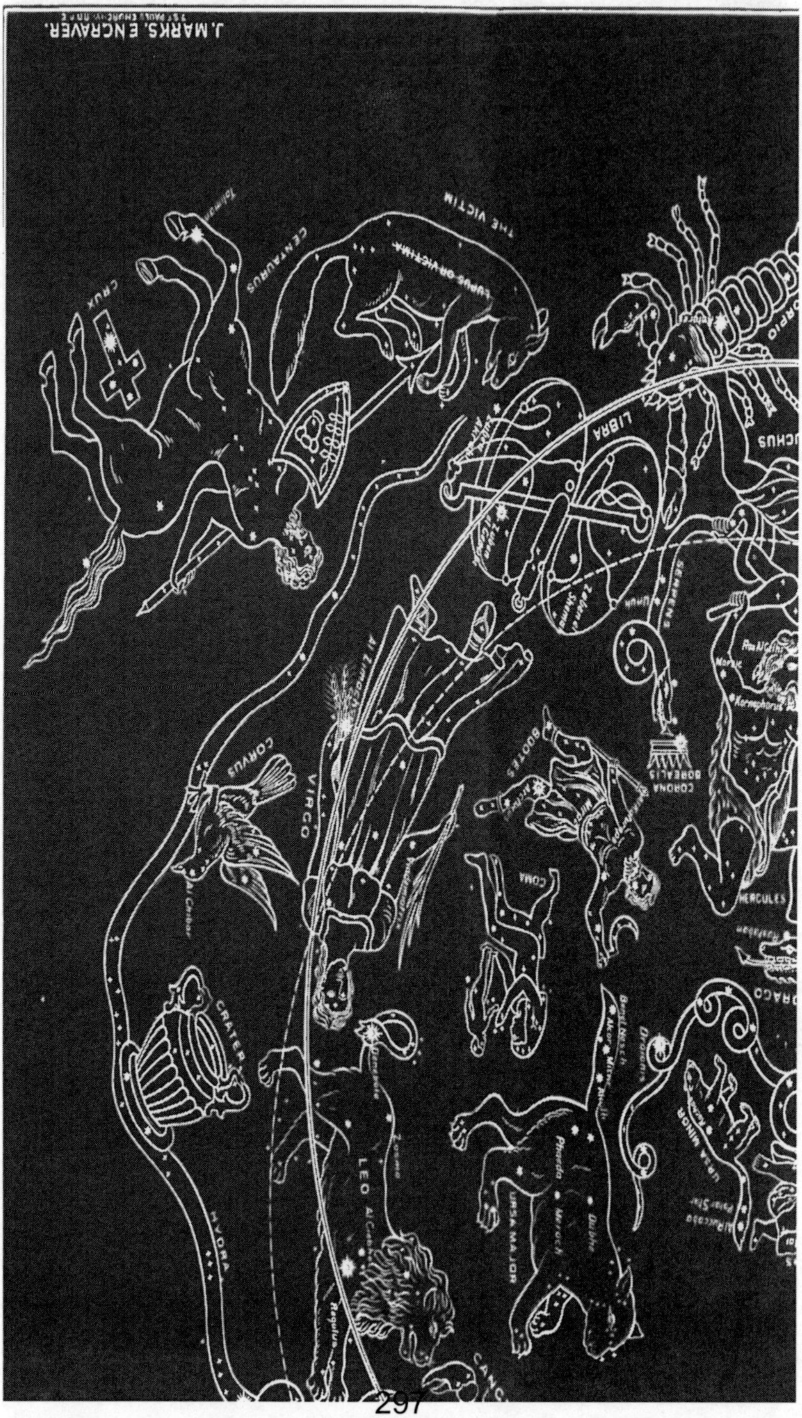

J. MARKS, ENGRAVER.
CRUX
CENTAURUS
THE VICTIM
LUPUS OR VICTIMA
LIBRA
SERPENS
CORVUS
VIRGO
CRATER
HYDRA
LEO
Regulus
BOOTES
CORONA BOREALIS
COMA
HERCULES
URSA MAJOR
URSA MINOR

ANDROMEDA
PERSEUS
CASSIOPEIA
ECLIPTIC
AQUARIUS
PEGASUS
CEPHEUS
Al Rai
Al Ruccaba
Polar Star
URSA MINOR
Al Deramin
PICIS AUSTRALIS
CYGNUS
Deneb
DRACO
Rastaban
DELPHINUS
SAGITTA
CAPRICORNUS
THE EQUATOR OF THE HEAVENS
LYRA
HERCULES
Al Tair
Tarazed
AQUILA
Ras al Hague
OPHIUCHUS
SAGITTARIUS
THE ECLIPTIC OR THE SUN'S PATH
SCORPIO
Antares
STARS OF THE FIRST MAGNITUDE
" " " SECOND "
" " " THIRD "
" " " FOURTH "
" " " FIFTH "
" " " SIXTH "
ARA

WORKS BY THE REV. E. W. BULLINGER, D.D.

FOUR PROPHETIC PERIODS.

A Key to the "Things which must Shortly come to Pass."

"MAN'S DAY." 1 *Cor. iv.* 2.
"THE DAY OF CHRIST." *Phil. ii.* 16.
"THE DAY OF THE LORD." *Rev. i.* 10.
"THE DAY OF GOD." 2 *Pet. iii.* 12.

Third Edition. Price One Penny.

"THE MAN OF GOD:"

A Bible Study.

Price One Penny: Seven and Sixpence per 100. Second Edition.

"THINGS TO COME:"

An Epitome of Prophetic Truth.

Envelope size. Price One Penny: Seven and Sixpence per 100.
Third Edition.

"THE KINGDOM" AND "THE CHURCH";

Or, The Seven Parables of Matthew xiii.

Price One Penny: Seven and Sixpence per 100. Second Edition.

THE NEW CREATION AND THE OLD;

Or, The Ways of God in Grace, illustrated by the Works of God in Creation.
A Bible Study on Genesis i. in the light of 2 Cor. v. 17.

Price One Penny: Seven and Sixpence per 100. Third Edition.

SUNDAY SCHOOL LESSONS.

A Series for Five Years in Course of Publication.

Monthly (Single Lessons), One Penny each. Yearly Volumes, bound, One Shilling. FIRST and SECOND Series now ready.

THE INSPIRATION AND AUTHORITY OF HOLY SCRIPTURE.

Third Edition. Price One Penny: Seven and Sixpence per 100.

GOD'S PURPOSE IN ISRAEL:

Third Edition, Revised. Price One Penny: Seven and Sixpence per 100.

HYMNS ON THE SECOND ADVENT.

A Collection of 118 Hymns for use in Conferences, &c., on the Subject of the Pre-Millennial Advent of Christ.

Price Sixpence, cloth, gilt lettered. Special terms for quantities.

Published by the Author, 7, St. Paul's Churchyard, London, E.C.

WORKS BY THE REV. E. W. BULLINGER, D.D.

THE WITNESS OF THE STARS;

Showing how the Ancient Signs and Constellations were designed for "signs and seasons" to record and preserve the prophecy of Genesis iii. 15 for 2,500 years before the Revelation from God was written in the Scriptures of Truth.

With more than 41 coloured plates and other engravings. Handsomely bound, suitable for presents. Price Seven and Sixpence.

NUMBER IN SCRIPTURE;

ITS SUPERNATURAL DESIGN AND SPIRITUAL SIGNIFICANCE.

Demy 8vo., Illustrated. Price Five Shillings.

THE STRUCTURE OF THE TWO EPISTLES TO THE THESSALONIANS,

Arranged after the manner of the *Key to the Psalms*, on two large sheets folded in envelope. Price Sixpence.

CHRIST'S PROPHETIC TEACHING

IN RELATION TO THE DIVINE ORDER OF HIS WORDS AND WORKS.

Demy 8vo., paper covers, 32 pages. Price Sixpence. Second Edition.

THE RESURRECTION OF THE BODY.

Being an Address at the Mildmay Conference, 1893.

Price One Penny : Seven and Sixpence per 100.

HYMNS FOR BIBLE READINGS.

Thirty-five Hymns.

Price One Penny : Seven and Sixpence per 100.

FIFTY ORIGINAL HYMN-TUNES.

Set to Appropriate Words.

Second Edition. Price Two Shillings, post free.

SIXTY-SIX OLD BRETON TUNES.

Collected and transcribed from the lips of Breton peasants ; and harmonised in short score, for the use of the churches of the Breton Evangelical Mission. These Tunes have never before appeared in this form, and most of them are printed for the first time. Some of them are in the ancient Greek modes, and all of them are peculiarly characteristic.

Price Two Shillings, post free.

Published by the Author, 7, St. Paul's Churchyard, London, E.C.